ZERO POINTS OF VEDIC ASTRONOMY

Also by the author:

1. Vedic Physics: Scientific Origin of Hinduism
2. India before Alexander: A New Chronology
3. India after Alexander: The Age of Vikramādityas
4. India after Vikramāditya: The Melting Pot
5. Zero Point of Jain Astronomy: The Origin of Malava Era

Zero Points of Vedic Astronomy

Discovery of the Original Boundaries of Nakshatras

Raja Ram Mohan Roy, Ph.D.

Mount Meru Publishing

Library and Archives Canada Cataloguing in Publication

Title: Zero points of Vedic astronomy : discovery of the original boundaries of Nakṣatras / Raja Ram Mohan Roy, Ph.D.
Names: Roy, Raja Ram Mohan, 1966- author.
Description: Includes bibliographical references and index.
Identifiers: Canadiana (print) 20190147105 | Canadiana (ebook) 20190147164 | ISBN 9781988207247
(softcover) | ISBN 9781988207230 (HTML)
Subjects: LCSH: Astronomy—India—History. | LCSH: Hindu astronomy.
Classification: LCC QB18 .R69 2020 | DDC 520.954—dc23

Published in 2020 by:
Mount Meru publishing
P.O. Box 30026, Cityside Postal Outlet PO
Mississauga, Ontario, Canada L4Z 0B6
Email: mountmerupublishing@gmail.com
Web: https://www.mountmerupublishing.com/
Facebook: https://www.facebook.com/MountMeruPublishing
ISBN 978-1-988207-24-7

Dedicated to Varāhamihira

CONTENTS

PREFACE

It was 20 years ago that I became aware of the problem of the dating of Varāhamihira. According to Indian tradition, Varāhamihira was one of the nine gems in the court of Emperor Vikramāditya, the greatest hero of ancient India. Emperor Vikramāditya is associated with the Vikrama era, which has its zero point in 57 BCE. Modern history denies the existence of Emperor Vikramāditya in 57 BCE and Varāhamihira has been placed in sixth century CE. When did Varāhamihira live, first century BCE or sixth century CE? This question has occupied my mind since then and resulted in reexamining the very foundations of Indian history. After a lot of research, I have come to the conclusion that the foundations of Indian history are wrong. I have presented my reconstruction of Indian history in my books India before Alexander: A New Chronology, India after Alexander: The Age of Vikramādityas, and India after Vikramāditya: The Melting Pot. In these books I have reinterpreted the existing evidence to construct an alternative timeline of Indian history.

During my investigation of the dating of Varāhamihira, I came to know that Varāhamihira has specified the position of nakṣatras during summer and winter solstices. Currently these positions are assumed to have been true during sixth century CE as that is the time assigned to Varāhamihira. However, if Varāhamihira lived during first century BCE, then the position of nakṣatras specified by him would have been true in first century BCE. This would mean that currently believed nakṣatra boundaries are off by approximately 10°. I have investigated this problem for last two years in detail. The findings of my research in archaeoastronomy carried out for this purpose are being presented in this book and a companion to this book titled "Zero Point of Jain Astronomy: The Origin of Mālava Era."

I would like to thank Mr. Mithilesh Jha and Mr. Gauri Shankar Jha for reading the draft and providing comments and suggestions. I would also like to thank Dr. Kamal Nath Jha for the pictures of Ashoka pillar in Prayagraj. Thanks are also due to Dr. Manish Mehta and Dr. Sulekh Jain for their support and encouragement. I would like to express my sincere appreciation to my wife Manju for her continued and enthusiastic support for this work.

Raja Ram Mohan Roy
Mississauga, Canada
January 15, 2020

"Wrong does not cease to be wrong because the majority share in it."
— Leo Tolstoy

1. Flawed Sheet Anchors of Indian History

The history of India as written in textbooks was compiled during the time India was colonized by Britishers. Before that Indians believed in a history as described in the Rāmāyaṇa, Mahābhārata, and Purāṇas. European scholars worked out the chronology of Indian history by identifying the connections between Indian and European historical figures. They identified two sheet anchors that firmly tie the Indian history to Greek history. Most of the ancient Indian history has been constructed by counting backward and forward from these sheet anchors as shown in Figure 1.1.

1.1 First sheet anchor of Indian history

The first sheet anchor is the identification of Sandrokottos of the Greek accounts with Chandragupta Maurya, the founder of the Mauryan Dynasty. It was in 1793 CE, when Sir William Jones, President of the Asiatic Society, made this discovery public:

> This discovery led to another of greater moment; for Chandragupta, who, from a military adventurer, became, like Sandracottus, the sovereign of Upper Hindustan, actually fixed the seat of his empire at Pataliputra, where he received ambassadors from foreign princes; and was no other than that very Sandracottus who concluded a treaty with Seleucus Nicator; … [1]

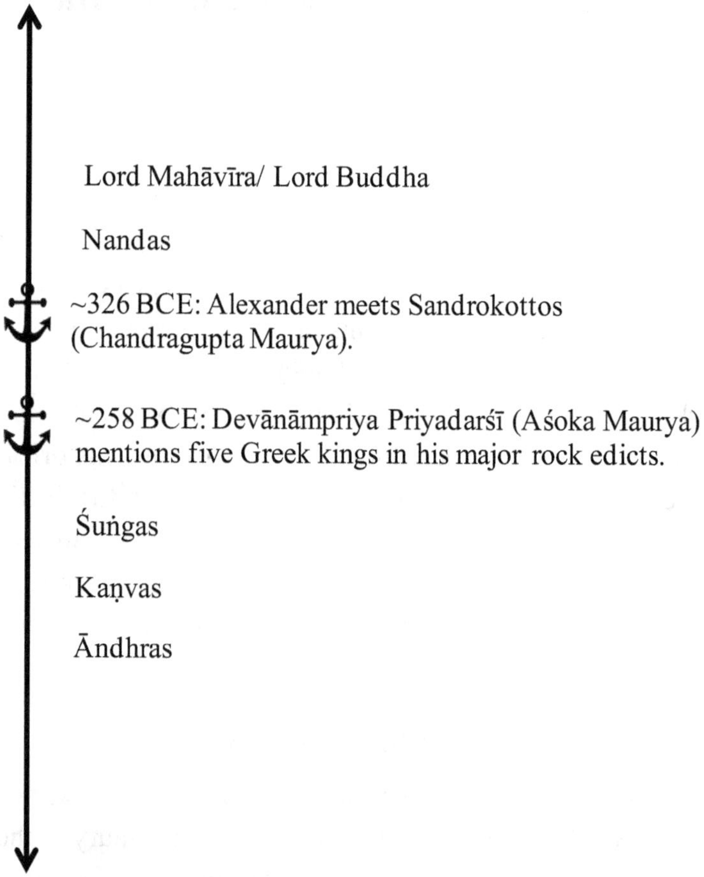

Figure 1.1: The construction of Indian chronology from accepted sheet anchors

J.W. McCrindle, compiler of several books detailing the ancient Greek writings on India, described this discovery in following words:

> The discovery that the Sandrokottos of the Greeks was identical with the Chandragupta who figures in the Sanskrit annals and the Sanskrit drama was one of great moment, as it was the means of connecting Greek with Sanskrit literature, and of thereby supplying for the first time a date to early Indian history, which had not a single chronological landmark of its own. [2]

However, the identification of Chandragupta Maurya, the founder of the Mauryan Dynasty, with Sandrokottos is not unique. There is another namesake, Chandragupta I of the Imperial Gupta Dynasty, who could also be the Sandrokottos of the Greek accounts. Currently, Chandragupta Maurya is considered the contemporary of Alexander the Great and Seleucus I Nicator. What if, Chandragupta I of the Imperial Gupta Dynasty was the contemporary of Alexander the Great and Seleucus I Nicator? Ancient Indian history will then be off by more than six centuries. Since the Greek accounts only give the phonetic equivalent of first name Chandragupta and not the last name Maurya, either Chandragupta Maurya or Chandragupta I of the Imperial Gupta Dynasty could be meant by them. It is then important to examine rest of the evidence to see which Chandragupta fits the Greek accounts better.

Greek classical writers Diodorus and Curtius have named Xandrames [3] or Agrammes [4] as the ruler of India before Sandrokottos, respectively. There is no phonetic similarity between Xandrames or Agrammes and Dhana Nanda, the ruler of India before Chandragupta Maurya. Greek classical writers have named Amitrochates [5] or Allitrochades [6] as the ruler of India after Sandrokottos. The names Allitrochades and Amitrochates have no similarity with Bindusāra, son and successor of Chandragupta Maurya. According to history books, Bindusara used the title Amitraghāta ("Slayer of enemies"), the phonetic equivalent of Amitrochates. There is no evidence to this effect. This is a deduction based on the identification of Sandrokottos with Chandragupta Maurya, but it is taught like there is some independent inscriptional or literary evidence to this effect. A better phonetic equivalent to Amitrochates is the Sanskrit term "Amitrochchhetā" ("Mower of enemies"), which is similar to the term "Sarvarājochchhetā" ("Mower of all kings") applied to Samudragupta, son of Chandragupta I, by his successors [7].

Greek classical writer Strabo has written that Seleucus I Nicator gave land gift to Sandrokottos in concluding a marriage alliance, and received in exchange 500 elephants [8]. The details of marriage are not given, but modern historians assume that Seleucus gave his daughter in marriage to Chandragupta Maurya due to his defeat in the war. It is entirely possible that Seleucus gave his daughter in marriage to Samudragupta, son of Chandragupta I. Samudragupta has claimed in the Eran Stone Inscription [9] to have earned his wife Dattadevī using his prowess. This will make perfect sense for a marriage alliance as a result of success in war. The first part of the name "Datta" means "given", as in given due to defeat in war. The second part Devī is simply an honourable name for a woman. Thus Dattadevī could be the name given to the daughter of

Seleucus after marriage to Samudragupta. Sandrokottos had an encounter with Alexander prior to Alexander's war with King Porus in 326 BCE. The war between Sandrokottos and Seleucus took place in circa 304 BCE. As Chandragupta would be a middle-aged man at this time and Samudragupta a young man, it is plausible that the daughter of Seleucus was married to Samudragupta.

There are other pieces of evidence described in my book "India before Alexander: A new Chronology" [10] that show that Chandragupta I of the Imperial Gupta Dynasty better matches the Sandrokottos of the Greek accounts instead of Chandragupta Maurya. However, what tilts the balance in favour of Chandragupta Maurya being Sandrokottos is the second sheet anchor of Indian history.

1.2 Second sheet anchor of Indian history

The second sheet anchor of Indian history is the identification of Devānāmpriya Priyadarśī of major rock edicts with Aśoka Maurya, the grandson of Chandragupta Maurya. It was Prinsep, who identified Devānāmpriya Priyadarśī with Aśoka Maurya in 1838:

> "Mr. Turnour fixes the date of Aśoka's accession in B.C. 247, or 62 years subsequent to Chandragupta, the contemporary of Seleucus. Many of his edicts are dated in the 28th year, that is in B.C. 219, or six years after Antiochus the Great had mounted the throne." [11]

In the thirteenth rock edict of Devānāmpriya Priyadarśī, five Greek rulers are mentioned who are currently assumed to be his contemporaries:

> And this (conquest) has been won repeatedly by Devanampriya both here and among all (his) borderers, even as far as at (the distance of) six hundred yojanas, where the Yona king named Antiyoka (is ruling), and beyond this Antiyoka, (where) four-4-kings (are ruling), (viz. the king) named Turamaya, (the king) named Antikini, (the king) named Maka, (and the king) named Alikasudara, (and) towards the south, (where) the Choḍas and Pāṇḍyas (are ruling), as far as Tāmraparṇī. [12]

This is the piece of evidence on which the chronology of Indian history rests. Five Greek kings mentioned by Devānāmpriya Priyadarśī are Antiyoka, Turamaya, Antikini, Maka and Alikasudara. Their phonetic equivalents are Antiochus, Ptolemy, Antigonus, Magas, and Alexander respectively. Antiyoka is currently identified as Antiochus II. However, Prinsep had initially identified Antiyoka as Antiochus III the Great [11], but later changed the identification to Antiochus I or II [13]. This is of major significance in the correct identification of Devānāmpriya Priyadarśī of major rock edicts.

Although it has been known for long that there is an alternative identification possible for the first sheet anchor of Indian history as discussed earlier in this chapter, no one has come up with a reasonable alternative identification for the second sheet anchor of Indian history. An alternative explanation was given by Somayajulu, as quoted by Pandit Kota Venkatachelam. According to him, Devānāmpriya Priyadarśī was not Aśoka Maurya but Aśokāditya, which was another name of Samudragupta:

> The so-called inscriptions of Aśoka do not belong to Aśoka. Most of them do not make any mention of Aśoka. If one or two mention Aśoka they do not refer to Aśoka Vardhana of the Maurya dynasty but they refer to Samudragupta of the Gupta dynasty who assumed the title of Aśokaditya. [14]

The problem is that there is no evidence whatsoever that Samudragupta ever took the title Aśokaditya, and there is no match between the characters of Samudragupta and Devānāmpriya Priyadarśī. However, the characters of Aśoka Maurya and Devānāmpriya Priyadarśī don't match either. So blinded are historians by the minor rock edicts identifying Devānāmpriya Priyadarśī as Aśoka that they have never bothered to think that the existence of Devānāmpriya Priyadarśī Aśoka does not mean that every Devānāmpriya Priyadarśī was Aśoka. The chronology of Indian history has been fixed by the identification of Devānāmpriya Priyadarśī of major rock edicts with Aśoka Maurya, but nowhere in the major rock edicts does Devānāmpriya Priyadarśī call himself Aśoka. This is very important since major rock edicts name the five Greek kings linking Indian history to Greek history, while minor rock edicts don't name Greek kings. Thus the contemporaneity of Aśoka and Greek kings is not directly established and depends on the Devānāmpriya Priyadarśī of minor rock edicts being same as the Devānāmpriya Priyadarśī of major rock edicts.

We can find out if the Devānāmpriya Priyadarśī Aśoka of minor rock edicts was same as the Devānāmpriya Priyadarśī of major rock edicts by comparing Devānāmpriya Priyadarśī known from major rock edicts with Aśoka Maurya known from literature. The information from the major rock edicts must match the information from literature if both are the same.

There is plenty of literary information available about Aśoka Maurya in: 1. Chronicles of Sri Lanka; 2. Aśokāvadāna as preserved in Divyāvadāna and Chinese versions; 3. Records of Chinese pilgrims; 4. Rājataraṅgiṇī of Kalhaṇa; and 5. Purāṇas [15]. The chronicles of Sri Lanka include Dipavansha and Mahavansha, while records of Chinese pilgrims include travel notes of Fa-Hien and Yuan Xang. The 14 major rock edicts of Devānāmpriya Priyadarśī are given in the "Inscriptions of Aśoka" by Hultzsch [12]. A comparison of Aśoka Maurya from literature with Devānāmpriya Priyadarśī from major rock edicts is presented below.

The Conquest of Kaliṅga

According to Rock edict 13, the conquest of Kaliṅga and the remorse from the ravages of war were the most important events in the life of Devānāmpriya Priyadarśī, but these events find no mention in the literature about Aśoka Maurya. The Kaliṅga war was the turning point in the life of Devānāmpriya Priyadarśī. After the Kaliṅga war Devānāmpriya Priyadarśī decided to change his ways and he accepted Buddhism. But literary sources about Aśoka Maurya are totally silent about the Kaliṅga war. A.L. Basham, author of "The Wonder that was India", has noted this in his paper on Aśoka and Buddhism [15].

The conversion to Buddhism

According to Rock edict 13, the Kaliṅga war was the main factor behind the conversion of Devānāmpriya Priyadarśī to

Buddhism. However, according to Theravada tradition, Aśoka Maurya was converted by a seven-year-old monk with no relation to the Kaliṅga war [15]. According to Fa-Hien, Aśoka was converted by a Buddhist monk, who was being tortured by Aśoka [16], again with no relation to the Kaliṅga war. There is no mention in literature that Aśoka Maurya converted to Buddhism due to the Kaliṅga war.

Third Buddhist Council

According to Buddhist literary sources, the Third Buddhist Council was held under the patronage of Aśoka Maurya, but there is no mention of it in the edicts of Devānāmpriya Priyadarśī. The absence is very glaring, as Devānāmpriya Priyadarśī describes matters of far less significance in his edicts about what he has done to promote Dharma.

The Family

Aśoka Maurya had sent his son Mahendra and daughter Sanghamitra to Sri Lanka to spread Buddhism. There is no mention of them in the edicts of Devānāmpriya Priyadarśī. From the inscription on the Allahabad Pillar, we know that Karuwaki was the wife of Devānāmpriya Priyadarśī, and Tivara was their son. However, both Karuwaki and Tivara are not mentioned in literary sources about Aśoka Maurya. In the fifth rock edict, Devānāmpriya Priyadarśī mentions his brothers and sisters, while according to Dipavansa and Mahavansa, Aśoka had killed all his 99 stepbrothers save his own brother Tissa. We have no mention of killing of step brothers in any of the inscriptions. Also, there is no mention of Tissa in any of his inscriptions. There is not a single person that is common to both literary sources about

Aśoka Maurya and inscriptions of Devānāmpriya Priyadarśī. Mahendra and Sanghamitrā from literature are not mentioned in the inscriptions, and Karuwaki and Tivara from the inscriptions of Devānāmpriya Priyadarśī are not to be found in the literature about Aśoka Maurya. The identification of Karuwaki as wife of Aśoka Maurya and Tivara as their son is based on the identification of Aśoka Maurya with Devānāmpriya Priyadarśī of Allahabad pillar, which may not be true.

Vegetarianism

Aśoka was a Jain before conversion to Buddhism according to Rājataraṅgiṇī 1.101-102. Chandragupta Maurya, grandfather of Aśoka Maurya, was a Jain who had spent the latter days of his life serving the Jain saint Bhadrabahu. Aśoka's grandson Samprati was also a Jain. So if Aśoka's grandfather was a devout Jain and his grandson Samprati was a devout Jain, it is natural to assume that Aśoka Maurya was also born a Jain. As Jains and Buddhists are both vegetarians, Aśoka was a vegetarian before and after conversion to Buddhism. However, Devānāmpriya Priyadarśī says in his major rock edicts that before his conversion hundreds of thousands of animals were killed daily in the royal kitchen. This is incompatible with Aśoka always being a vegetarian, first as a Jain and then as a Buddhist. Devānāmpriya Priyadarśī must have been a meat-eating Hindu before becoming a Buddhist.

Tolerance

Aśoka, who is considered an apostle of non-violence, was not so tolerant even after his conversion to Buddhism.

According to Aśokavadana, once Ājīvikas made a painting showing Buddha as subordinate to the founder of the Ājīvika sect. Aśoka was enraged and he ordered all the Ājīvikas of Pundravardhana (North Bengal) to be killed. Eighteen thousand Ājīvikas lost their lives in just one day [17-18]. Devānāmpriya Priyadarśī followed non-violence after his conversion to Buddhism according to his inscriptions, and it would be out-of-character for him to have ordered the massacre of Ājīvikas.

These arguments show that the identification of Devānāmpriya Priyadarśī of major rock edicts with Aśoka Maurya is not as sacrosanct as the modern historians would make us believe. The question then is, "who was Devānāmpriya Priyadarśī of major rock edicts"? Is there any other candidate for identification as Devānāmpriya Priyadarśī, who will fit the available evidence better?

Before I present an alternative identification for Devānāmpriya Priyadarśī, let us assume for argument's sake that historians have made a mistake in identifying Sandrokottos with Chandragupta Maurya, who should really be identified with Chandragupta-I of Imperial Gupta dynasty. Chandragupta Maurya is currently assumed to have ruled during the last quarter of fourth century BCE, while Chandragupta-I is supposed to have ruled during the first half of fourth century CE. The two Chandraguptas are separated in time by roughly 650 years. If Chandragupta-I has been shifted forward in time by over six centuries, then a lot of evidence would need to be twisted to fit the currently accepted chronology. This is indeed the case as discussed in the next chapter.

Notes

1. Jones (1793): xii-xiv.
2. McCrindle (1877): Footnote on page 7.
3. McCrindle (1893): 281-282.
4. McCrindle (1893): 221-222.
5. Hamilton (1892): 109.
6. McCrindle (1893): 409.
7. Sethna (1989): 246.
8. McCrindle (1901): 88-89.
9. Fleet (1888): 20-21.
10. Roy (2015a).
11. Prinsep (1838a).
12. Hultzsch (1925): 27-71.
13. Prinsep (1838b).
14. Venkatachelam (1953): 8.
15. Basham (1982).
16. Legge (1886): 90-92.
17. Mukhopadhyaya (1963): xxxvii.
18. Strong (1989): 232.

"If you never heal from what hurt you, you'll bleed on people who didn't cut you."

- Wanjiku Kamuyu

2. Force-fitting of Evidence in Indian History

Information about Indian history comes from many sources such as literature, inscriptions, numismatics and archaeology. Let's consider all this information, which comes in bits and pieces, as parts of a gigantic puzzle. If the framework of history is correct, all these pieces will fit together and fall in their proper places. On the other hand, if the framework of history is incorrect, historians will be force-fitting these puzzle pieces into places where they don't belong. Indian history is full of examples of this force-fitting.

Let's assume that Chandragupta-I of the Imperial Gupta dynasty was the contemporary of Alexander the Great instead of Chandragupta Maurya. This will imply that Imperial Gupta dynasty has been moved forward by over six centuries. This will have a cascading effect as shown in Figure 2.1.

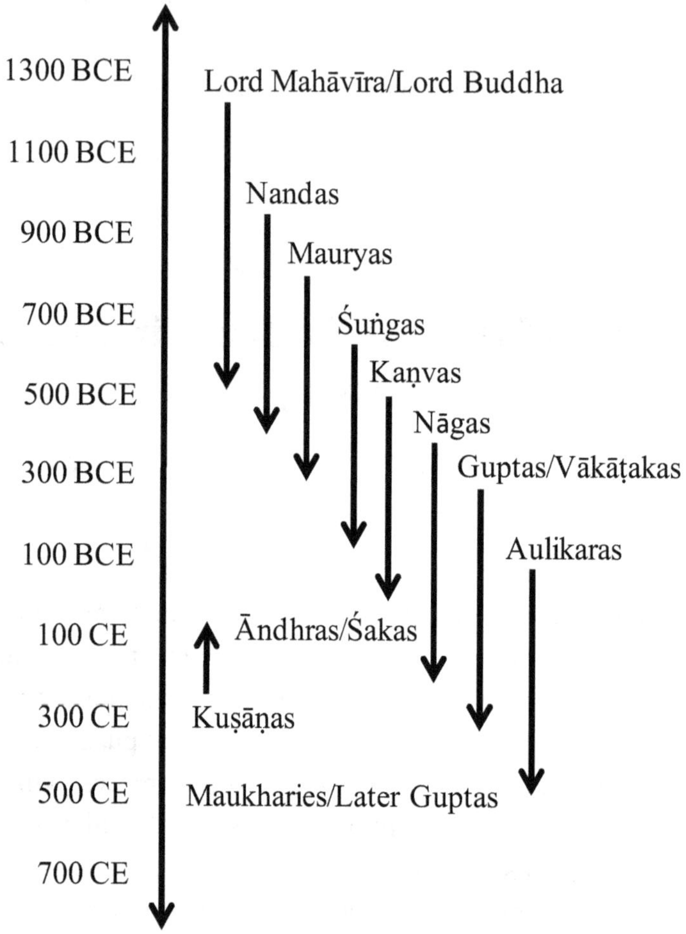

Figure 2.1: The consequences of chronology constructed from wrong sheet anchors

In this figure, a timeline has been drawn on the left side and different dynasties and important historical figures are shown on the right. The arrows qualitatively show their displacement from their actual time.

Let's discuss the consequences of chronology constructed from wrong sheet anchors as shown in Figure 2.1.

2.1 Telltale signs

1. An uninhabited time period

If Imperial Guptas have been moved forward by over six centuries, then dynasties before them have also been moved forward. Since historical time period is finite, this process will generate a big gap in the timeline. In fact, there is a big gap in Indian history spanning many centuries after the end of Indus Valley Civilization. An artificial theory called Aryan Invasion theory has been devised to fill this gap. If we look closely into the timeline, we find that the chronology before the Buddha is rather vague and amorphous. There are no names of any historical personages before the sixth century BCE. There is only literary history before this period without any names attached to any event. Thus, the people who are placed in the sixth century BCE may as well be placed in the twelfth century BCE without contradicting any evidence.

2. A congested time period

If the Imperial Guptas were moved forward by six centuries then those following the Guptas also needed to be moved forward accordingly. But obviously, the process will end somewhere as we cannot place ourselves six centuries ahead of our own time. In fact, the process has to end long before our time because we stand on much surer ground regarding the time period of historical events during the second millennium. This means that the timeline will get crowded where the people belonging to different eras will be made to face each other. This will result in a situation where history will declare certain people as contemporaries, even when there is no evidence to corroborate and support such contemporaneity. Is there any evidence of this phenomenon? I believe there is, and that time period is the sixth century CE.

Historians have proposed the existence of four mighty empires in the sixth century -- of the Imperial Guptas, the Aulikaras, the Maukharies, and the Later Guptas. While the end of Imperial Guptas coincided with the rise of the Aulikaras, and the Maukharies had marriage relationships with Later Guptas establishing their contemporaneity, there is no evidence that the first two were the contemporaries of the latter two. This has resulted in a situation where three different dynasties claim to have ruled North India within a short span of time:

> It should be remembered that thc time of both Iśānavarmā and Jīvitagupta I falls between 520 to 540 AD. This is exactly the time of Malwa's Aulikara emperor Yaśodharmā (known date 532 AD) and he has been given credit for winning the region from Himalayas in the north to Brahmaputra river to the east. Clearly he would have won Bengal as well. It is clear from this fact that Jivitagupta I, Crown prince Iśānavarmā and Malwa's Yaśodharmā, all three claim to have won Himalayan region and Bengal between 520 to 540 AD. [1]

This peculiar situation is a result of placing Aulikara emperor Yaśodharmā where he does not belong. His date of 532 CE is a direct result of equating Mālava era with Vikrama era as Yaśodharmā gives his date in Mālava era.

There is no evidence that Mālava era and Vikrama era were the same.

3. Future before past

There are many examples where an event happening earlier has been placed chronologically after an event happening later. The current sequence of these events is logically impossible. Here are three examples.

i. The world-renowned Nalanda University was established by Kumaragupta-I. According to accepted history, Kumaragupta-I reigned between 415 and 447 CE. Nāgārjuna, the famous Buddhist philosopher, is considered to have lived around 150 CE. Nāgārjuna not only studied at the Nalanda University, but also taught there. How did Nāgārjuna study and teach at the Nālandā University in 2nd century CE, if the university was founded in 5th century CE by Kumāragupta I?

ii. King Hāla Sātavāhana is currently dated in the first century CE. He has written about emperor Vikramāditya's generosity in a verse in Gāthāsaptaśatī 5.64. This is fatal to currently accepted chronology, as it does not accept any Vikramāditya in the first century BCE. The Imperial Gupta emperor it credits with being Vikramāditya is placed in the fourth century CE. How could Hāla Sātavāhana know about the generosity of Vikramāditya in first century, if Vikramāditya reigned in fourth century?

iii. The time of Vallabhī rulers is currently calculated based on the assumption that they were using the Vallabhī era with a starting date of 319 CE. The Alina copper plate inscription of Śilāditya VII was written in the year 447, which places the last Vallabhī ruler Śilāditya VII in 766 CE using the Vallabhī era [2]. According to numerous Rajput genealogies, the great Bappa Rawal was a direct descendant of Śilāditya (VII) and was separated from him by eight generations. This will make the great Bappa Rawal roughly 200 years posterior to Śilāditya VII. It is currently accepted that Bappa Rawal was born in 713 CE and died in 753 CE. So we have a situation where a descendent is chronologically placed before his ancestor. This has completely muddied the well-preserved traditions of Mewar and the genealogies maintained by the Rajputs.

4. False contemporaries

It was pointed earlier that Imperial Guptas and the Aulikaras have been falsely made contemporaries of the Maukharies and the Later Guptas. Another example of dynasties made contemporary is that of the Śakas and the Kuṣāṇas, this time because of moving Kuṣāṇas backward in time as shown in Figure 2.1. The necessity of this move is to avoid making Kuṣāṇas and Imperial Guptas contemporaries. This is impossible because Kuṣāṇas and Imperial Guptas ruled over domains that have large common territories. However, a consequence of making the

Śakas and the Kuṣāṇas contemporaries is that they both claim sovereignty over the same region at the same time:

> Rasesh Jamindar has advanced a theory that Kanishka existed after Rudradaman perhaps in the second half of the 2nd century. He argues that Rudradaman's domination over Sindhu, Sui Vihar (including Multan) and Kanishka's rule over Multan and Sui Vihar is not possible at the same time. Secondly Rudradaman could not have defeated Yaudheyas after crossing the Kushana territory who were occupying the Punjab area. [3]

Even though modern historians make the Śaka Kṣatrapas and Mahākṣatrapas the subordinates of Kuṣāṇa rulers, the fact is that the Kuṣāṇas have not mentioned the Śakas, and the Śakas do not know of any Kuṣāṇas. Similarly, we find some small rulers minting their own coins right under the nose of the mighty Imperial Guptas:

> The rule of Shakas and Shiladas came to an end in c. 340 A.D. with the rise of a tribe, which is sometimes described as the Little Kushana and sometimes as Kidara Kushana. ... Numismatic evidence shows that a number of petty rulers like Kritavirya, Shiladitya, Sarvayashas, Bhasvan, Kushala and Prakasha were ruling in the Punjab during the first half of the 5th century A.D. They were probably Kidara Kushana rulers, for the name Kidara appears on their coins on the obverse. [4]

There were other rulers also claiming sovereignty within the dominion of the Imperial Guptas at the very zenith of their power:

> Unlike the Maharajas of Valkha, Subandhu does not refer to any suzerain even in a general manner, which shows that he was an independent ruler. In 416-417 AC,

> the Gupta power had, no doubt, reached its peak. Chandragupta II was dead at the time and was succeeded by his son Kumaragupta I, but there is no reason to suppose that the Gupta dominion had suffered any diminution at the beginning of the latter's reign. It may, therefore, be asked how Kumaragupta allowed Subandhu to enjoy independence just on the border of the Avanti province which was undoubtedly under Gupta rule at the time. [5]

These examples should raise serious doubts about the accepted period of the Imperial Guptas' reign. Was it really between the fourth to sixth century CE?

In fact, there is strong archaeological evidence to the contrary. The age of the Imperial Guptas is considered the golden age of India. This was the age of unprecedented growth in prosperity, art and culture. However, the archaeological excavations present a completely different picture. In the period when the cities are supposed to be flourishing, there are definitive signs of rapid decay. Here is a summary of the status of Indian cities during the third to fifth century of the Common Era according to archaeological findings:

> After third century of Common Era, there was rapid decay in urban centres of Punjab, Haryana and western Uttar Pradesh, for example Hastināpura and Mathurā. During Gupta age, cities were fast deteriorating in middle Gangetic plains, for example Śrāvastī, Kauśāmbī and Rājagīra. Conditions were similar in West Bengal and Odisha. Urban decay started in North India with the rise of the Imperial Guptas. The signs of urban decay in cities administered by Guptas are clear in cities such as

Pāṭaliputra, Vaiśālī, Vārāṇasī, Kauśāmbī, Ayodhyā, Hastināpura and Mathurā. [6]

This contradicting evidence is due to the placement of Imperial Guptas where they don't belong.

5. Against established traditions

It is understandable if the traditional accounts are wrong about the hoary antiquities such as the date of Mahābhārata war in 4th millennium BCE, but can they be dead wrong about the events in 1st century BCE? The biggest example of the force-fitting of evidence in the making of Indian history is in regard to the tradition related to the Emperor Vikramāditya. According to Indian tradition, Emperor Vikramāditya was the greatest hero of ancient India. He had crushed the Śakas, who had dared to invade India. He was a paragon of virtue, his generosity was legendary, and he was a great patron of the arts. There were nine gems, people with extraordinary skills in their field, in his court. Two of these gems were Varāhamihira, astronomer par excellence, and Kālidāsa, poet par excellence. The Vikrama era, still in use in India, was instituted to honour Emperor Vikramāditya and has its zero point in 57 BCE. The love and respect that Emperor Vikramāditya commanded from Indians grew over time and he became the hero of a number of fables, as described in the Vetālapañchaviṃśati (popularly known as Vetāla Pachīsī), Siṃhāsana-dwātriṃśikā (popularly known as Siṃhāsana Battīsī), and Śuka-saptaśatī (popularly known as the story of a parrot and a mynah).

However, modern history denies the very existence of Emperor Vikramāditya in 57 BCE. Varāhamihira has been placed in sixth century CE based on the misinterpretation of the Śaka era. While Varāhamihira counted his time from the Cyrus Śaka era beginning in 550 BCE (as his forefathers had migrated to India from Persia in the wake of the devastation brought to Persia by Alexander), historians have simply refused to accept the existence of a Śaka era older than the Śalivāhana Śaka era beginning in 78 CE. Kālidāsa has been made the court poet of Chandragupta II Vikramāditya, who ruled between 376-415 CE. In the astrological text, Jyotirvidābharaṇa, supposedly written by Kālidāsa, the time of writing of this text is given as 3068 Kali, which is 34/33 BCE. This makes Kālidāsa a junior contemporary of Emperor Vikramāditya. There is no explanation for the Vikramāditya being in the fourth-fifth century CE, while the Vikrama era counts its beginning from 57 BCE.

6. Missing links

Ancient Indian and Persian civilizations were sister civilizations. There were interactions between the sister civilizations of India and Persia that have been denied by modern historians. According to Persian historians, a Persian emperor named Bahram came to India and married the daughter of Indian emperor Basdeo. There is not only literary but also numismatic proof to this effect from Persia as described below by Prinsep [7]:

> One confirmation of a historical fact from numismatic aid has been remarked in the discovery of the name of Vāsa Deva or Bas Deo on a Sassanian coin. Ferishta states, that Bas Deo, of Kannauj, gave his daughter in marriage to Behram of Persia, A.D. 330:- the coin marks exactly such an alliance; but the Hindu chronicles admit

no such name until, much later, one occurs in the Malwa catalogue of Abul Fazl.

Prinsep again mentions this marriage in the genealogical table of Kanauj where he says that Basdeo (Vasudeva) revived the Kanauj dynasty and his daughter married Bahram Sassan of Persia in 390 CE, according to Ferishtah [8]. Fergusson has made the following comments in 1870 CE regarding the coins with Vasudeva written on one side and an image of a Sassanian king on other side [9]:

> There is still another group of coins called Indo-Sassanian, which, however, have only been imperfectly read. The typical example of the class is one originally drawn by Prinsep, and produced by Thomas (vol. i, pl. vii., fig. 6.). It represents a Sassanian king on one side; on the other, another who may be an Indian with a distinctly legible inscription in Sanskrit characters, which reads Śrī Vasudeva. While the other inscriptions are undecyphered, it is too hazardous even to suggest that this may be the father-in-law of Bahram Gour; but the number of these Indo-Sassanian coins which are found in India, extending even beyond Hegira, prove a close intercourse between the two countries at the period we are now speaking about, and when thoroughly investigated, will, I fancy, throw more light on the political and religious changes that took place in India about the sixth century, than anything else which has yet come to light.

The possibility of the marriage of an Indian princess, daughter of King Vasudeva, with Persian emperor Bahram is denied by modern Indian historians. There was no king named Basdeo in India when Bahram ruled in Persia, according to these historians. However, this is the result of faulty chronology of Kuṣāṇa emperors. When the proper chronology is developed, this marriage alliance becomes chronologically feasible. The details of this chronological reconstruction are provided in my book "India after Vikramāditya: The Melting Pot" [10].

At this point, it will be pertinent to ask if the Indian history is so drastically wrong, won't this be obvious to historians. As discussed earlier, historians consider the identifications of Sandrokottos with Chandragupta Maurya and Devānāmpriya Priyadarśī with Aśoka Maurya as sacrosanct, and therefore have no option but to accept the chronology derived from these two identifications. Most of the work in developing this chronology was done by colonial era historians, some of whom had ulterior motives derived from their racist and imperial agendas. Here are the methods the colonial era historians have used to forcefully fit every bit of evidence in the currently accepted chronology.

2.2 Tricks of the trade

1. Leverage in choosing the zero points

None of the dates of any historical figure is based on scientific dating. If the dates are approximate, they are based on circumstantial evidence. If the dates are precise, they may be derived from a year given in an era or from a year counted from the establishment of a dynasty. In many cases the era given is ambiguous or not given, giving historians the flexibility to choose the era that does not contradict the established chronology. For example, the Imperial Guptas have been force-fitted in their current place in history by choosing a convenient zero point of

Imperial Gupta era, shifting the Kuṣāṇas backward in time, and shifting the Vallabhī and Gurjara dynasties forward in time. History books since the colonial times have been teaching that the Kuṣāṇa emperor Kaniṣka-I was the founder of the Śalivāhana Śaka era, which started in 78 CE. However, Kaniṣka was a Kuṣāṇa and not a Śaka. Modern historians have justified this by saying that the term Śaka was used by Indians for any foreigner. Professor Harry Falk has shown that according to the text Yavanajātaka by Sphujidhvaja, the Kuṣāṇa era started 149 years after the Śaka era, i.e. in 227 CE [11]. This means that currently the Kuṣāṇas occupy a place 149 years before their actual time.

2. Twisting the evidence to fit presumed chronology

The dates of Imperial Gupta emperors have been fixed by using a zero point of Imperial Gupta dynasty in 319 CE, but the evidence for this is doubtful. This date has been derived from the statement of Al-Biruni, who says the following:

> As regards the Guptakala, people say that the Guptas were wicked powerful people, and that when they ceased to exist this date was used as the epoch of an era. [12]

Al-Biruni clearly says that the Imperial Guptas had ceased to exist when the Imperial Gupta era commenced. This is what was originally used for fixing the date of Imperial Guptas as described on a signboard as shown in Figure 2.2. This signboard is next to the Allahabad Pillar in U.P., India and was placed in 1838 CE. The Allahabad pillar has inscription of Samudragupta, whose time is indicated as second century in the signboard.

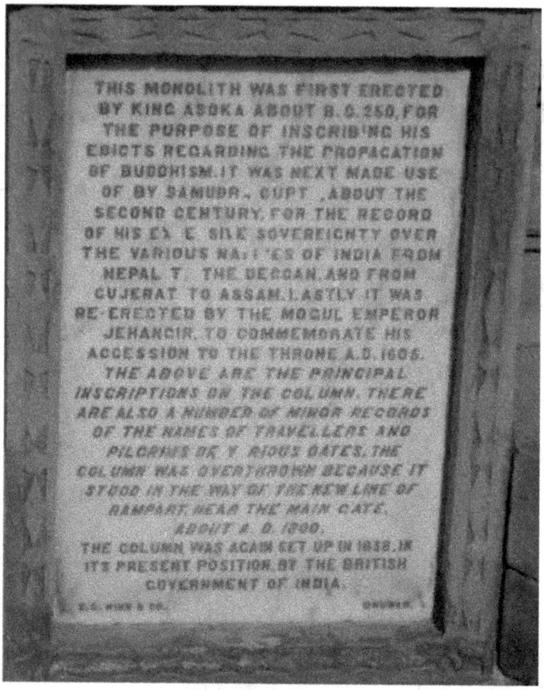
THIS MONOLITH WAS FIRST ERECTED
BY KING ASOKA ABOUT B.C.250, FOR
THE PURPOSE OF INSCRIBING HIS
EDICTS REGARDING THE PROPAGATION
OF BUDDHISM. IT WAS NEXT MADE USE
OF BY SAMUDR. GUPT, ABOUT THE
SECOND CENTURY, FOR THE RECORD
OF HIS SOVEREIGNTY OVER
THE VARIOUS OF INDIA FROM
NEPAL THE DECCAN, AND FROM
GUJERAT TO ASSAM. LASTLY IT WAS
RE-ERECTED BY THE MOGUL EMPEROR
JEHANGIR, TO COMMEMORATE HIS
ACCESSION TO THE THRONE A.D. 1605.
THE ABOVE ARE THE PRINCIPAL
INSCRIPTIONS ON THE COLUMN. THERE
ARE ALSO A NUMBER OF MINOR RECORDS
OF THE NAMES OF TRAVELLERS AND
PILGRIMS OF DATES. THE
COLUMN WAS OVERTHROWN BECAUSE IT
STOOD IN THE WAY OF THE NEW LINE OF
RAMPART NEAR THE MAIN GATE,
ABOUT A.D. 1800.
THE COLUMN WAS AGAIN SET UP IN 1838, IN
ITS PRESENT POSITION, BY THE BRITISH
GOVERNMENT OF INDIA.

Figure 2.2: A signboard next to Ashoka pillar in Prayagraj (photo courtesy: Dr. Kamal Nath Jha)

The date of second century was worked out in confirmation with the statement by Al-Biruni quoted above, which says that the Imperial Guptas had ceased to exist when the era in their name was started. However, historians later twisted the statement of Al-Biruni and used the zero point of Imperial Gupta era to commence the rule of Imperial Guptas and calculated the regnal years of Samudragupta to be from 350 to 376 CE. This example shows how conveniently historical people can be moved from one time period to another by twisting the evidence.

3. Pick and choose evidence

An example of the pick and choose approach is that of the dating of Gautama Buddha. When colonial era historians started to piece together the history of India, they considered the date of birth of Buddha to be 1027 BCE [13]. The date of birth of Buddha was revised subsequent to the identification of the Indian king Sandrokottos from Greek accounts with Chandragupta Maurya by Sir William Jones in 1793 CE [14]. Most of the modern historians place the birth of Buddha in the sixth century BCE (sometime between 567-563 BCE) and his death in the fifth century BCE (sometime between 487-483 BCE). Since Indian and Chinese dates are too early, modern historians have argued that Ceylonese/Sri Lankan dates are more reliable. It goes against common sense that the place farthest from the birthplace of Buddha would have preserved the most authentic date of his birth! The fact is that the Ceylonese texts "Dīpavaṃśa" and "Mahāvaṃśa" were written in the fourth century and fifth to sixth century respectively. These texts, in turn, are based on texts that are no longer available. There is simply no reason for Ceylonese texts to be more reliable than Indian, Chinese, and Nepalese texts.

Modern historians have calculated the date of the Buddha from the date of Aśoka Maurya. Since the date of coronation of Aśoka Maurya was fixed at ~268 BCE, based on his identification with Devānāmpriya Priyadarśī, historians searched for texts for the date of the Buddha that would be consistent with the date of Aśoka's coronation. They found in Ceylonese texts that coronation of Piyadassi took place 218 years after the death of the Buddha. Working backwards, historians calculated the date of the death of Buddha at ~486 BCE and his birth 80 years earlier at ~566 BCE. However, the same Ceylonese texts that mention 218 years between the death of Buddha and coronation of Piyadassi also say that the Buddha died in 544/543 BCE. If we take that date as reliable, then the coronation of Aśoka Maurya took place 218 years later in 326/325 BCE, which is around the time of the invasion of India by Alexander. This will make Aśoka Maurya the contemporary of Alexander instead of his grandfather Chandragupta Maurya. So we are being told by historians that Ceylonese texts are the most reliable regarding the dating of Gautama Buddha while these texts give the dates of Buddha's life and death that are completely irreconcilable. First historians are choosing Ceylonese texts to be more reliable than Indian, Chinese, and Nepalese texts, and then choosing one date to be reliable in the Ceylonese texts for calculating the date of Buddha, while rejecting the other date in the same texts without any valid reason.

4. <u>Declaring the evidence as forgery</u>

There are many examples of genuine evidence being declared forgeries because the evidence does not fit the established chronology. Here are three examples.

i. The details of Gokak copper plates were published by N. Lakshminarayana Rao in Epigraphia Indica [15]. These plates discovered in 1926 from a house in Gokak in the then Belgaum district of the Bombay Presidency mention an "Aguptāyika era" that has got historians totally perplexed. Rao, the author reporting the finding of Gokak plates, connected Aguptāyika era to Chandragupta Maurya [15]. However, the argument was found unconvincing by modern historians, as an era related to Chandragupta Maurya would be named after Mauryas, not Guptas. So we have the following statement from noted historian D.C. Sircar about the Aguptayika era:

> Aguptayika era may be roughly assigned to 845-645 = 200 B.C. … But we can scarcely accept the evidence of a single inscription regarding the existence of a genuine era starting from about 200 B.C. in the face of the overwhelming negative evidence. … The story may have been fabricated by the astronomer at Dejja-Maharaja's court. [16]

Here we have an evidence of declaring genuine evidence as fabrication. Nobody fabricates this kind of stuff as there is no motivation for it. It is a proof that the rule of Imperial Guptas tooks place long before currently held time period.

ii. An astronomical text called the Sumatitantra is the first book on astronomy from Nepal. There is a research paper on Sumatitantra titled "Mānadeva Samvat: An investigation into a Historical Fraud" by Kamal P. Malla [17]. We should note the title of the paper, which is symptomatic of the attitude that modern historians have towards our ancient records. The objective is not to understand what they mean but to declare as forgery whatever does not suit the accepted chronology. What is being dismissed as a historical fraud not only provides information about the date of Buddha, but also provides evidence in support of the Imperial Guptas being contemporary of Alexander. In addition, it provides the identification of Emperor Śudraka, whose writing Mṛichchhakaṭikam is well known (Film Utsava was based on it), but whose identity is unknown. The details are provided in my book "India after Alexander: The Age of Vikramadityas" [18].

iii. The time of the Gurjaras is currently calculated based on the assumption that they were using Kalchuri/Chedi era with a starting date of ~248 CE. However, there is documentary evidence of the existence of three inscriptions of the Gurjaras in which the Śaka era has been used. These inscriptions occur on three plates identified as Bagumrā, Ilāo and Umetā plates by Bühler in a paper written in 1888 CE [19]. These inscriptions were declared forgeries by colonial era historians as they contradicted the established timeline. When we take into account the evidence of Bagumrā, Ilāo and Umetā plates, we find that Gurjara rulers have been shifted forward from their actual time by about 170 years. The details of its effect on the chronology of Indian history are provided in my book "India after Vikramāditya: The Melting Pot" [20].

5. Selective destruction of evidence

As discussed above, the Bagumrā, Ilāo and Umetā plates were declared forgeries by colonial era historians. Not only that, their existence itself has been erased. The two parts of Corpus Inscriptionum Indicarum, Volume 4 [21-22] are supposed to list and present all inscriptions belonging to Early Gurjara rulers, whether considered genuine or forged, but the inscriptions of Bagumrā, Ilāo and Umetā plates are not found in this volume. I have found the existence of these plates, only because Google has digitized the journals and books from colonial era and placed them in public domain.

Also, many inscriptions of the Vallabhī rulers have gone missing. Colonial era scholar Bhandarkar had written the following in 1872:

> Dr. Bhau Dāji gives, in one place, the dates of five copper plate grants of this dynasty, whilst in another he mentions seven dates professedly derived from copper plates. But he does not say when or by whom so many grants of the Vallabhī kings were discovered, nor who deciphered and translated them, or where the plates of their transcripts and translations are to be found. [23]

It stands to reason that all inscriptions were vetted by British authorities and only those inscriptions have survived which in the eyes of the colonial authorities did not directly contradict the official chronology.

6. Making up innovative theories without evidence

An example of this is the preposterous claim that an itsy-bitsy ruler named Azes was the founder of the Vikrama era:

> Azes (Aya in Kharosthi) was another powerful Śaka ruler in the Northwest who initiated a dynastic era beginning in 58/57 B.C. which later became identified with the so-called Vikrama era still used in South Asia. [24]

This petty ruler Azes was a Śaka ruler, whereas, according to Indian tradition, Emperor Vikramāditya is known as Śakāri – the enemy of the Śakas! We might as well ask our modern historians why the era of Azes is not called the Azes era. Why would it be called the Vikrama era? This strange and inexplicable sleight of hand by colonial era historians is an untenable act of a deliberate manipulation of historical records. Now this lie has been exposed as Falk and Bennet have shown that the Azes era did not start in 57 BCE [25].

With the material presented in this chapter, I believe that I have given sufficient evidence to seriously doubt the accepted chronology of Indian history. I will now present an alternative chronology which I believe fits the evidence better.

Notes

1. Goyala (1986): 23-24 (translated from Hindi).
2. Fleet (1888): 171-191.
3. Sagar (1992): 171.
4. Majumdar and Altekar (1967): 21-23.
5. Mirashi (1955a): xxxix-xl.
6. Śarmā, R. (1995): 225-227 (translated from Hindi).
7. Thomas (1858): 221.
8. Thomas (1858): 258.
9. Fergusson (1870).
10. Roy (2015c): 28-48.
11. Falk (2001).
12. Sachau (1910): 7.
13. Dietz (1995): 39-105.
14. Jones (1793).
15. Rao (1931-32).
16. Sircar (1965): 326.
17. Malla (2005).
18. Roy (2015b): 64-72.
19. Bühler (1888).
20. Roy (2015c):76-99.
21. Mirashi (1955a).
22. Mirashi (1955b).
23. Bhandarkar (1872).
24. Srinivasan (2007): 71.
25. Falk and Bennett (2009).

"As long as one can think as an outsider, an observer apart from the conflict, there is hope for a resolving thought."
— R. N. Prasher

3. Alternative Sheet Anchors of Indian History

There are many glaring inconsistencies in the currently accepted version of Indian history which have resulted from the flawed sheet anchors of Indian history. Most of ancient Indian historical chronology has been constructed by counting backward and forward from the two sheet anchors of Indian history. The key to discovering the true history of India is to correctly identify the Devānāmpriya Priyadarśī of major rock edicts.

If the identification of Devānāmpriya Priyadarśī of major rock edicts with Aśoka Maurya is wrong, then who was the real Devānāmpriya Priyadarśī of the major rock edicts? Obviously, he is well known to historians due to the extent of his vast empire. He is just not known to historians as the Devānāmpriya Priyadarśī. My personal research on this topic started in 2001 after I realized that the currently accepted dating of Varāhamihira in sixth century CE is wrong. This dating is based on using Śālivāhana Śaka era starting in 78 CE while Varāhamihira has specified his date using Cyrus Śaka era starting in 550 BCE [1]. According to my calculation, an astronomical observation made by Varāhamihira places him in the 2nd century BCE. This makes it possible for Varāhamihira to be a senior contemporary of Emperor Vikramāditya. This raised the possibility that there was a historical Vikramāditya in 57 BCE as the Vikrama era is counted from 57 BCE. Historians have denied the possibility of any Vikramāditya in first century BCE and have given credit for instituting the Vikram era to an itsy-bitsy ruler called Azes. This petty ruler was a Śaka, while Indian tradition considers Vikramāditya to be Śakāri or enemy of Śakas. Not only that, Emperor Vikramāditya was a ruler of all of India and beyond according to the traditions, while the dominion of this petty ruler Azes was much smaller and not worthy of any special recognition. I then searched for a historical Vikramāditya and realized that Emperor Yaśodharmā was the historical Emperor Vikramāditya. However, historians have placed him in sixth century CE based on the wrong zero point of Mālava era. Since the time of Imperial Guptas was before the time of Emperor Yaśodharmā, the timing of Imperial Guptas needed to be fixed. This could only be done by re-evaluating the very foundations of Indian history. So my work should be viewed as my modest attempt to get the greatest hero of ancient India, Vikramāditya Yaśodharmā the Great, his due place in history.

When I was searching through history books for the alternative identification of Devānāmpriya Priyadarśī, I found the history books written by Shriram Goyala very useful. Though his books conform to the accepted Indian history, he provides lot of background information that I needed for my own analysis. In one of his books, he describes a coin which shows the central figure as a Buddhist monk, who is possibly being implored by two other figures in the coin not to become a monk. The emperor shown as a Buddhist monk is Kumāragupta-I, and it seemed to me that he is possibly the Devānāmpriya Priyadarśī of major rock edicts. With further research, I became convinced that the identification of Kumāragupta-I as the Devānāmpriya Priyadarśī of major rock edicts is the key to discovering the true history of India. Now I will provide my reasoning for this identification by comparing what we know of Kumāragupta-I, great grandson of Chandragupta-I of the Imperial Gupta Dynasty, with Aśoka Maurya, the currently accepted Devānāmpriya Priyadarśī of major rock edicts.

3.1 The Kaliṅga War

According to Rock edict 13, the conquest of Kaliṅga and the remorse from the ravages of war were the most important events in the life of Devānāmpriya Priyadarśī. However, these events find no mention in the literature about Aśoka Maurya. On the other hand, there is literary evidence that Kaliṅga was conquered by Kumāragupta-I. The following text from Viṣṇupurāṇa (4.24.64-65) describes the expansion of the Imperial Gupta Empire:

> Kośala Oḍratāmraliptān Samudrataṭa Purīm cha Rakṣito Rakṣyati|
> Kaliṅgam Māhiṣakam Mahendraḥ Bhūmau Guham Bhokṣyanti|| [2]

Śrīrāma Goyala explains the meaning of this verse as follows:

> (Deva) Rakṣita will expand his domain to Kośala, Oḍra, Tāmralipti and Purī near ocean. Kaliṅga and Māhiṣaka will be under Mahendra. All this land will be ruled by Guha. [2]

Here Rakṣita stands for Gupta, as both those words mean "protected", and Mahendra stands for Kumāragupta-I Mahendrāditya. Guha stands for Skandagupta, as Guha and Skanda are synonyms. This important verse gives the following information:

> Gupta (Chandragupta II) will protect the territories of (South) Kośala, Oḍra, Tāmralipti, Samataṭa and Purī (which are already part of the Gupta Empire). Kumāragupta-I will expand it further to include Kaliṅga and Māhiṣaka. Skandagupta will enjoy ruling all this land.

Here, we have emphatic proof that Kaliṅga was not a part of the Gupta Empire ruled by Chandragupta II, but was conquered by Kumāragupta-I. This is the war that changed Kumāragupta-I, and he accepted Buddhism soon after. Compare this to Aśoka Maurya for whom we have no independent information that he had to fight a war to incorporate Kaliṅga into his empire. In fact, the evidence points to the opposite. Aśoka should have inherited Kaliṅga as it was part of the Nanda Empire, which was taken over by his grandfather Chandragupta Maurya in a coup. Aśoka did not have to fight a war to capture Kaliṅga. To circumvent this problem, modern historians have made up a story about Kaliṅga gaining independence from the Mauryan Empire before the coronation of Aśoka. There is absolutely no evidence to this effect. In fact, there is evidence that indicates that Kaliṅga could not have seceded from Mauryan Empire before Aśoka. Chāṇakya is supposed to have served three kings -- Chandragupta, Bindusāra and Aśoka -- according to the medieval text ĀryaManjuśrīMūlakalpa [3]. It would have been very unlikely for Kalinga to secede under the watch of Chāṇakya.

3.2 Junagadh Rock Inscription

Skandagupta, son of Kumāragupta-I, says the following in the line four of the Junagadh rock inscription [4]:

"Pitari sura-sakhitvam prāptvaty ātma- śaktyā"

The meaning of each word in this sentence is provided below:

Pitari = father, sura = Gods, sakhitvam = friendship, prāptvaty = obtain, ātma = self and śaktyā = from power

Thus the sentence means that the father obtained the friendship of the Gods by his own power. Fleet has translated it as "father by his own power had attained the position of being a friend of the gods" [4]. Historians have taken it to mean that Kumāragupta-I had passed away when this inscription was recorded, as it is customary in India to say that a person has become dear to God when he or she has passed away. Fine, but how did Kumāragupta-I do it with his own power? Did he commit suicide? We don't have any record of that and if he did commit suicide, why would his son Skandagupta be proudly announcing it? What the sentence in the inscription actually means is that Kumāragupta-I had obtained the friendship of the Gods by his own power while he was still alive. At least that is what his son Skandagupta was made to believe as his father

Kumāragupta-I had declared himself "Beloved of the Gods" in inscriptions all over the vast empire. Skandagupta was just paraphrasing the word "Devānāmpriya", meaning "Beloved of the Gods" to "Friend of the Gods".

3.3 Man of Many Names

If someone calls himself "Devānāmpriya" and "Priyadarśī", besides his own name, then we can definitely call him a person with many names. There is evidence that Kumāragupta-I was known as a man with many names. ĀryaMañjuśrīMūlakalpa is a Sanskrit text written by a Buddhist around 800 CE. It was translated into English by noted historian K. P. Jayaswal. This text gives the following information about the Imperial Guptas:

> Listen about the Medieval and Madhyadesa kings (madhyakāle, madhyamā) who will be in a long period emperors (nṛpendra) and who will be confident and will be followers of via media" (in religious policy, madhyadharmiṇaḥ):
> (1) Samudra, the king,
> (2) Vikrama, of good fame (kīrttitāḥ), 'who is sung'.
> (3) Mahendra, an excellent king and a leader (nṛpavaro Mukhya).
> (4) S-initialled (Skanda) after Ma. (i.e., Mahendra).
> His name (will be) Devarāja; he will have several names (vividhākhya); he will be the best, wise, and religious king in that low age. [5]

Above, the first king is Samudragupta; the second king is Chandragupta II, referred to by the first part of his title Vikramāditya; the third king is Kumāragupta-I, referred to by the first part of his title, Mahendrāditya; and, the fourth king is Skandagupta, identified by his initial S. In the passage quoted above, I would like to draw the readers' attention to the description of the king called Devarāja, who was supposed to have several names. Jayaswal has identified him with Skandagupta [6]. Jayaswal says that Skandagupta bore the name of his grandfather (Devarāja), and had a variety of names (virudas). There is no evidence that Skandagupta bore the name of Devarāja after his grandfather.

It is my contention that Devarāja refers to Kumāragupta-I and hence it indicates that Kumāragupta-I was known by many names. The author says "S-initialled (Skanda) after Ma" and then goes on to say "His name (will be) Devarāja". This raises the possibility that Devarāja refers to the king with initial M, i.e. Mahendrāditya, adopted name of Kumāragupta-I. Devarāja means King of the Gods, which is Indra. Kumāragupta-I has been called Mahendra by the author of ĀryaMañjuśrīMūlakalpa, as quoted above. Mahendra (Mahā + Indra) is simply "Great Indra" or "Indra himself", and thus it is Kumāragupta-I who has been called Devarāja and a man of many names. This indirectly corroborates Kumāragupta-I being the Devānāmpriya Priyadarśī of major rock edicts.

3.4 The Samudragupta Pillar

There are several sets of inscriptions on the Ashoka Pillar currently located in the Allahabad Fort, including inscriptions by Devānāmpriya Priyadarśī, his queen, and most importantly, Samudragupta the Great. A picture of this pillar is shown in Figure 3.1. The evidence of this pillar is so significant that it is enough to invalidate the currently accepted version of Indian history.

Figure 3.1: The Samudragupa pillar in Prayagraj (photo courtesy: Dr. Kamal Nath Jha)

The inscription by Samudragupta is considered posterior to the inscription by Devānāmpriya Priyadarśī based on his identification with Aśoka Maurya. Thus according to historians, Samudragupta got his eulogy inscribed on an existing pillar that already had the inscriptions of Aśoka Maurya. My contention is that pillar was originally erected by Samudragupta and the inscription of Devānāmpriya Priyadarśī was inscribed on it later.

Samudragupta was among the greatest conquerors known to Indian history. His eulogy inscribed on this pillar gives the details of his conquests and expanse of his empire. Samudragupta was known for the reestablishment of traditional Vedic/Hindu way of life. There is simply no reason for Samudragupta to have his eulogy inscribed on an existing pillar with inscriptions by a Buddhist king as his zeal for military conquests did not match the pacifist ideology of Devānāmpriya Priyadarśī. Thapar, wondering why Samudragupta chose to write his eulogy on the Aśokan pillar, says that extolling military conquest was contradictory to Aśoka's opposition to violence, and if Samudragupta wanted to denigrate Aśoka it would have been more effective on a separate and equally imposing pillar [7].

It is inconceivable that such a great monarch as Samudragupta, whose generosity was legendary, would use Aśoka Maurya's pillar for writing his eulogy. According to the inscription on this pillar, Samudragupta was so generous that he gave away hundred thousand cows (line 25). He called himself the God of Wealth, Kubera (line 26). He further said that his officials were busy returning the wealth of defeated kings everyday (line 26). Why would such a monarch not be able to afford a pillar of his own and choose a pillar erected by the Buddhist monarch Aśoka Maurya to write his eulogy? Not only that, why would he call somebody else's pillar a symbol of his glory? In the lines 29-30 of his inscription on this pillar, Samudragupta says with pride that this pillar is looking towards the heaven as the declaration of his glory [8]. On the other hand, Devānāmpriya Priyadarśī, identified as Kumāragupta-I by me, would have been more than happy to add his inscriptions on Samudragupta's pillar as his proud grandson. **If my identification is correct, then Ashoka pillars are in fact Kumāragupta pillars.**

3.5 A common title

The name Aśoka appears in a few minor rocks edicts as "Devānām Piya Aśoka" at Maski in Raichur district, Karnataka, as "Rājā Aśoko Devānāmpiya" at Udegolam in Bellary district, Karnataka, and as "Devānāmpiya Piyadasi Aśoka Rājā" at Gujarra near Jhansi, Madhya Pradesh [9]. Based on these inscriptions, the Devānāmpriya Priyadarśī of major rock edicts is identified as Aśoka Maurya. However, Devānāmpriya and Priyadarśī were common titles that could be used by anyone who chose to do so. Just because these titles have been used by Aśoka does not mean that nobody else could use these titles. When Prinsep was translating the inscriptions of Priyadarśī, he identified Priyadarśī first with Devānāmpriya Tissa of Ceylon [10-11].

The title Devānāmpriya has been used for other personalities in literature as well. King Ajātaśatru has been called "Devanuppiya" in "Aupapatika Sūtra". Patañjali, commenting on Pāṇini's Aṣṭādhyāyī 2.4.56, has used this title for a common grammarian. Priyadarśī or Priyadarśana can have two meanings: one who looks handsome or one who looks with friendliness. Priyadarśī was an adjective that has been used for several kings. In the Rāmāyaṇa, Rāma has been called Priyadarśī once. In the play, Mudrārākṣasa, Chandragupta Maurya, grandfather of Aśoka Maurya, has been called Priyadarśī. Gautamīputra Śātakarṇi has been called Priyadarśana in the Nasika inscription. There is nothing unique about the titles Devānāmpriya and Priyadarśī.

3.6 The identity of Antiyoka

As discussed, the chronology of Indian history rests on the contemporaneity of Aśoka Maurya and the five Greek kings Antiyoka, Turamaya, Antikini, Maka, and Alikasudara mentioned in the Rock Edict XIII. Modern historians have identified them with Antiochus II Theos (261-246 BCE) of Syria and Western Asia, Ptolemy II Philadelphus (285-247 BCE) of Egypt, Antigonus Gonatas (278-239 BCE) of Macedonia, Magas (300-258 or 250 BCE) of Cyrene, and Alexander (275-255 BCE) of Epirus or Alexander (252-247 BCE) of Corinth respectively [12]. Based on this information, historians have been able to pinpoint the date of coronation of Aśoka to within a couple of years:

> The latest date at which these kings were reigning together is 258, the earliest 261; and if we could be certain that Aśoka was kept informed of what happened in the West, we might therefore fix the twelfth year of his reign between these two years; and hence the date of his coronation between 270 and 273 B.C. [13]

Here is the relevant text of Rock Edict XIII:

> Antiyoke nāma Yona Rāja paran cha tena
> Antiyokena chatura rājāne Turamaye nāma
> Antikini nāma Maka nāma Alikasandare nāma [14]

The text has the following meaning: "The Greek king named Antiyoka, and beyond that king Antiyoka, four kings, named Turamaya, named Antikini, named Maka, and named Alikasandara". It is obvious that Devānāmpriya Priyadarśī had close interaction with Antiyoka or Antiochus and he probably had just heard about the other four Greek kings. When Prinsep first identified Antiyoka, he had identified him with Antiochus III and not Antiochus II as done by current historians. This is of critical importance as Antiochus II was involved in constant warfare and the connections between Mesopotamia and the borderlands of India were entirely cut off during his entire rule.

> What interests us in this connection is, however, not so much the character of Antiochus II as the main events of his reign. He undoubtedly inherited from his father a war with Egypt, which came to an end only during his very last years, and an unbroken series of troubles with the petty despots and quarrelsome city-states of Asia Minor. As far as the very scanty evidence goes, Antiochus II spent the whole of his reign in the last-named country and in Syria; and there is certainly no evidence whatsoever for his having ever proceeded to the east of the Mesopotamian rivers to visit the outlying provinces of his vast and loosely-knitted empire. Furthermore, we have the direct evidence of the historians, above all that of Justin, the epitomator Pompei Trogi, that during the reign of Antiochus II the most important provinces of the east rebelled, an event which must have entirely cut off the connections between Mesopotamia and the borderlands of India until these were again, for a very short period of time, restored by Antiochus the Great. [15]

On the other hand, identification of Antiyoka with Antiochus III the Great is on solid grounds as he came to the border of India and was thus known to Devānāmpriya Priyadarśī:

> The first point to be adjusted is, which Antiochus is referred to. There are several of the names amongst the

> kings of the Seleucidan dynasty, whose sway commencing in Syria, extended at various times, in the early periods of their history, through Persia to the confines of India. Of these, the two first, Antiochus Soter and Antiochus Theos, were too much taken up with concurrences in Greece and in the west of Asia, to maintain any intimate connexion with India, and it is not until the time of Antiochus the Great, the fifth Seleucid monarch, that we have any positive indication of an intercourse between India and Syria. It is recorded of this prince that he invaded India, and formed an alliance with its sovereign, named by Greek writers, Sophagasenas. [16]

If Antiochus III the Great was the contemporary of Devānāmpriya Priyadarśī, then Devānāmpriya Priyadarśī cannot be Aśoka Maurya as Antiochus III the Great reigned from 222 to 191 BCE, which will be too late for Aśoka Maurya as grandson of Chandragupta Maurya, who met Alexander the Great in 326 BCE. However, this is fine for Devānāmpriya Priyadarśī Kumāragupta-I as the great grandson of Chandraragupta-I.

The identifications of Sandrokottos with Chandragupta-I and Devānāmpriya Priyadarśī with Kumāragupta-I, both of the Imperial Gupta dynasty, provide us with alternative sheet anchors to reconstruct the history of India as shown in Figure 3.2. This reconstruction has been presented in detail in my books India before Alexander: A new Chronology, India after Alexander: The age of Vikramādityas, and India after Vikramāditya: The Melting Pot [17-19]. These books cover the time period from 13th century BCE to 7th century CE and present an alternative timeline of Indian history vastly different from the accepted timeline.

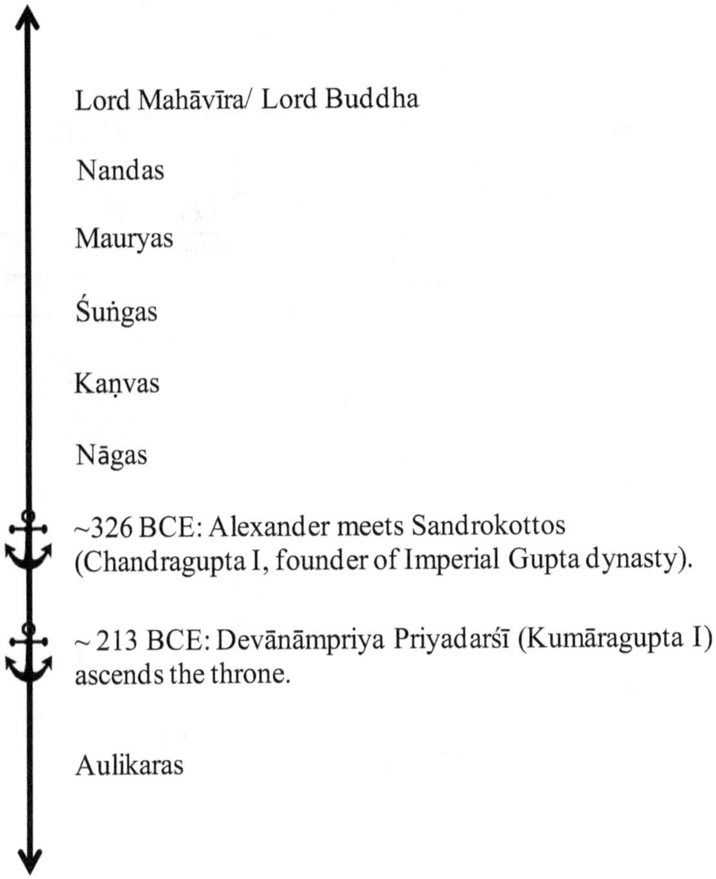

Figure 3.2: Construction of chronology based on alternate sheet anchors

Based on my work, a comparison of accepted dates of historical figures and dynasties with alternative dates is shown in Table 3.1. Most of the historical figures and dynasties have been moved forward by over six centuries. However, Kuṣāṇa dynasty has been moved backward to make room for Imperial Guptas.

Table 3.1: Accepted and alternate chronologies

	Accepted Chronology	Alternate Chronology [17-19]
Lord Mahāvīra	599-527 BCE or 540-468 BCE	1244-1172 BCE
Lord Buddha	563-483 BCE	1258-1178 BCE
Nandas	344-323 BCE	1019-919 BCE
Chandragupta Maurya	324-300 BCE	919-895 BCE
Bindusāra Maurya	300-273 BCE	895-870 BCE
Aśoka Maurya	273-236 BCE	870-834 BCE
Śuṅga dynasty	188-76 BCE	696-576 BCE
Kaṇva dynasty	76-31 BCE	576-531 BCE
Chandragupta I	319-50 CE	309-294 BCE
Samudragupta	350-76 CE	294-252 BCE
Chandragupta II	376-415 CE	252-213 BCE
Kumāragupta I	415-447 CE	213-173 BCE
Skandagupta	456-467 CE	172-161 BCE
Prakāśadharmā	515 CE	130 BCE
Yaśodharmā	532 CE	113 BCE
Kuṣāṇa dynasty	78 CE-320 CE	227-475 CE
Vallabhī dynasty	502-766 CE	144-543 CE

Currently there is a huge gap in Indian history after the fall of Indus Valley civilization. No such gap exists based on the correct identification of sheet anchors of Indian history. Table 3.1 summarizes the chronology of Indian history from 13th century BCE to 6th century CE. The history of India from 17th century BCE to 13th century BCE will be presented in my forthcoming book "India before Buddha: Vedic Kingdoms in 2nd Millennium BCE".

The contemporaneity of Imperial Gupta dynasty with Alexander the Great and Seleucus I Nicator is of major significance in the history of science, especially astronomy. It should be noted that major advances in astronomy in Greece took place after Alexander's invasion and subsequent contact between Greece and India. Colonial era historians had a vested interest in showing that the flow of knowledge was from Greece to India, as a proof of their supremacy. To achieve this objective, mere manipulation of historical evidence was not enough. They also needed to manipulate the meaning of astronomical texts, as astronomical observations can be dated. As the history of India has been pushed forward by over 600 years, the dates of astronomical observations have also been pushed forward by over 600 years. To understand how this manipulation has been effected, we will begin with the most fundamental concept of Indian astronomy, that of the nakṣatras.

Notes:

1. Roy (2015a).
2. Goyala (1987a): 253.
3. Jayaswal (1934): 17.
4. Fleet (1888): 56-65.
5. Jayaswal (1934): 33.
6. Jayaswal (1934): 35-36.
7. Thapar (2013): 341.
8. Goyala (1987b):16-19.
9. Vassilkov (1997-98).
10. Prinsep (1837).
11. Hultzsch (1914).
12. Sethna (1989): 233.
13. Davids (1877): 42.
14. Hultzsch (1925): 86-87.
15. Charpentier (1931).
16. Wilson (1850).
17. Roy (2015a).
18. Roy (2015b).
19. Roy (2015c).

"I shall now speak of the knowledge of the Hindus ... of their subtle discoveries in the science of astronomy – discoveries even more ingenious than those of the Greeks and Babylonians – of their rational system of mathematics, or of their method of calculation which no words can praise enough – I mean the system of nine symbols."

- Severus Sebokht

4. The Concept of Nakṣatras

Vedic people had been carefully observing the sky since the dawn of civilization. A critical examination of these observations has the potential to provide us with vital clues about the very early period of our history. The precession of the equinoxes is a well known astronomical phenomenon. It is related to the wobbling of the earth's axis and has a period of about 26,000 years. When we link the information on the precession of the equinoxes with observed positions of the nakṣatras as described in the ancient texts, we can obtain information about the time when these observations were made.

It takes the moon ~27.32 days to return to the same position among the stars. Based on this measurement, the path of moon in the background of the stars was divided in 28 or 27 divisions, each division being called a nakṣatra or lunar mansion. Atharvaveda Saṃhitā (19.7.1-5) lists the 28 nakṣatras as follows:

> 1. Seeking favour of the twenty-eight fold wondrous ones, shining in the sky together, ever-moving, hasting in the creation (Bhūvana), I worship (sapary) with songs the days, the firmament (nāka).
> 2. Easy of invocation for me [be] the Kṛttikās and Rohiṇī; be Mṛgaśirās excellent, [and] Ārdrā healthful (Śām); be the two Punarvasus pleasantness, Pushya what is agreeable, the Āśleṣās light (Bhānu), the Maghās progress (āyana) [for me].
> 3. Be the former Phālgunīs and Hasta here auspicious (puṇyam); be Chitrā propitious, and Svāti easy (sukhā) for me; be the two Viśākhās bestowal (rādhas), Anurādhā easy of invocation, Jyeshṭhā a good asterism, Mūla uninjured.
> 4. Let the former Ashāḍhās give me food; let the latter ones bring refreshment; let Abhijit give me what is auspicious; let Śrāvaṇa [and] the Śravishṭhās make good prosperity.
> 5. Let Śatabhishaj [bring] to me what is great widely; let the double Proshṭhapadas [bring] to me good protection (suśarman]; let Revatī and the two Aśvayuj [bring] fortune to me; let the Bharaṇīs bring to me wealth. [1]

The Abhijit nakṣatra was later dropped and the system of 27 nakṣatras became standard. There is a dialogue between gods Indra and Skanda regarding the dropping of nakṣatra Abhijit in Mahābhārata (Vana Parva, 230: 8-10). In this dialogue, Indra says to Skanda that because of jealousy with Rohiṇī, her younger sister Abhijit has gone to forest to do penance. This is a figurative way of saying that Abhijit has been dropped from the list of nakṣatras. There was also a change in the yogatārā of Revatī nakṣatra. This is figuratively told in the story of dropping of Revatī and her reinstatement as described in Chapter 72 of Mārkaṇḍeya Puarāṇa.

Each nakṣatra was assigned a presiding deity or set of deities. The list of 27 nakṣatras with their deity/deities is given in the Taittirīya Saṃhitā (iv.4.10) as follows:

> (Thou art) Krittikas, the Naksatra, Agni, the deity; ye are the radiances of Agni, of Prajapati, of the creator, of Soma; to the Re thee, to radiance thee, to the shining thee, to the blaze thee, to the light thee.
>
> (Thou art) Rohini the Naksatra, Prajapati the deity; Mrigaśirsa the Naksatra, Soma the deity; Ardra the Naksatra, Rudra the deity; the two Punarvasus the Naksatra, Aditi the deity; Tisya the Naksatra, Brihaspati the deity; the Aśresas the Naksatra, the serpents the deity; the Maghas the Naksatra, the fathers the deity; the two Phalgunis the Naksatra, Aryaman the deity; the two Phalgunis the Naksatra, Bhaga the deity; Hasta the Naksatra, Savitr the deity; Chitra the Naksatra, Indra the deity; Svati the Naksatra, Vayu the deity; the two Viśakhas the Naksatra, Indra and Agni the deity; Anuradha the Naksatra, Mitra the deity; Rohini the Naksatra, Indra the deity; the two Viśrits the Naksatra; the fathers the deity; the Asadhas the Naksatra, the waters the deity; the Asadhas the Naksatra, the All-gods the deity; Śrona the Naksatra, Visnu the deity; Śravistha the Naksatra, the Vasus the deity; Śatabhisaj the Naksatra, Indra the deity; Prosthapadas the Naksatra, the goat of one foot the deity; the Prosthapadas the Naksatra, the serpent of the deep the deity; Revati the Naksatra, Pusan the deity; the two Aśvayujs the Naksatra, the Aśvins the deity; the Apabharanis the Naksatra, Yama the deity. [2]

Earliest known astronomical text of India is Vedāṅga Jyotiṣa. It gives the list of the presiding deities of 27 nakṣatras [3] and thus indirectly provides the list of the 27 nakṣatras. Table 4.1 lists the nakṣatras and presiding deities as given in Atharvaveda Saṃhitā, Taittirīya Saṃhitā and Vedāṅga Jyotiṣa.

It can be seen that the main difference between the lists of nakṣatras in Atharvaveda Saṃhitā and Taittirīya Saṃhitā is the dropping of Abhijit nakṣatra in Taittirīya Saṃhitā. Some of the nakṣatras have slightly different names in the lists. The lists of presiding deities are nearly same in Taittirīya Saṃhitā and Vedāṅga Jyotiṣa and thus we can infer that the lists of nakṣatras in both texts were identical. Taittirīya Brāhmaṇa (1.5.1.1-5) also lists the 27 nakṣatra with the presiding deities. The list of nakṣatras started with Kṛttikā and ended with Bharaṇī in all these texts.

Taittirīya Brāhmaṇa (1.5.2.7) divides the list of nakṣatras in two groups. First group is called Deva nakṣatra and consists of 14 nakṣatras from Kṛttikā to Viśākhā. Second group is called Yama nakṣatra and consists of 13 nakṣatras from Anuradha to Bharaṇī. Deva nakṣatra means the nakṣatras belonging to the gods and Yama nakṣatra means the nakṣatras belonging to Yama, god of death in Hindu mythology. Some nakṣatras had alternative names: Puṣya (Tiṣya), Śravishṭhā (Dhanishṭhā), Pūrva-Proṣṭhapadā (Pūrva-Bhādrapadā), Uttara- Proṣṭhapadā (Uttara-Bhādrapadā), and Aśvayuja (Aśvinī).

Table 4.1: The list of nakṣatras

	Nakṣatras		Nakṣatras	Deities	Deities
	Atharvaveda Saṃhitā [3]		Taittirīya Saṃhitā [4]		Vedāṅga Jyotiṣa [5]
1.	Kṛttikā	1.	Kṛttikā	Agni	Agni
2.	Rohiṇī	2.	Rohiṇī	Prajāpati	Prajāpati
3.	Mṛgaśirā	3.	Mṛgaśīrṣa	Soma	Soma
4.	Ārdrā	4.	Ārdrā	Rudra	Rudra
5.	Punarvasu	5.	Punarvasu	Aditi	Aditi
6.	Puṣya	6.	Tiṣhya	Bṛhaspati	Bṛhaspati
7.	Āśleṣā	7.	Āśreshā	Serpents	Serpents
8.	Maghā	8.	Maghā	Fathers	Fathers
9.	Pūrva-Phālgunī	9.	(Pūrva) Phālgunī	Aryamā	Bhaga
10.	Uttara-Phālgunī	10.	(Uttara) Phālgunī	Bhaga	Aryaman
11.	Hasta	11.	Hasta	Savitā	Savitā
12.	Citrā	12.	Citrā	Indra	Tvaṣṭā
13.	Svāti	13.	Svātī	Vayu	Vayu
14.	Viśākhā	14.	Viśākhā	Indra and Agni	Indra and Agni
15.	Anurādhā	15.	Anurādhā	Mitra	Mitra
16.	Jyeṣṭhā	16.	Rohiṇī	Indra	Indra
17.	Mūla	17.	Vichṛta	Fathers	Nirṛti
18.	Pūrvāṣāḍhā	18.	(Pūrva) Āshāḍhā	Waters	Waters
19.	Uttarāṣāḍhā	19.	(Uttara) Āshāḍhā	All-gods	All-gods
20.	Abhijit				
21.	Śravaṇa	20.	Śroṇa	Viṣṇu	Viṣṇu
22.	Śraviṣṭhā	21.	Śraviṣṭhā	Vasu	Vasu
23.	Śatabhiṣaja	22.	Śatabhiṣaja	Indra	Varuṇa
24.	Pūrva-Proshṭhapadā	23.	(Pūrva) Proshṭhapadā	Goat of one foot	Goat of one foot
25.	Uttara-Proshṭhapadā	24.	(Uttara) Proshṭhapadā	Serpent of the deep	Serpent of the deep
26.	Revatī	25.	Revatī	Pushā	Pushan
27.	Aśvayuja	26.	Aśvayuja	Aśvins	Aśvins
28.	Bharaṇī	27.	Apabharaṇī	Yama	Yama

Figure 4.1 illustrates the principle of the division of celestial sphere in nakṣatra zones. K and K′ represent north ecliptic pole and south ecliptic pole respectively. A, B, C, and D are the boundaries of nakṣatras on the ecliptic. In a 27 nakṣatra system, there will be 27 such points on the ecliptic through which the boundaries of nakṣatras will pass. As the nakṣatras have equal span in the 27 nakṣatra system according to Sūrya Siddhānta 2.64, each nakṣatra has a span of 13° 20′ on the ecliptic.

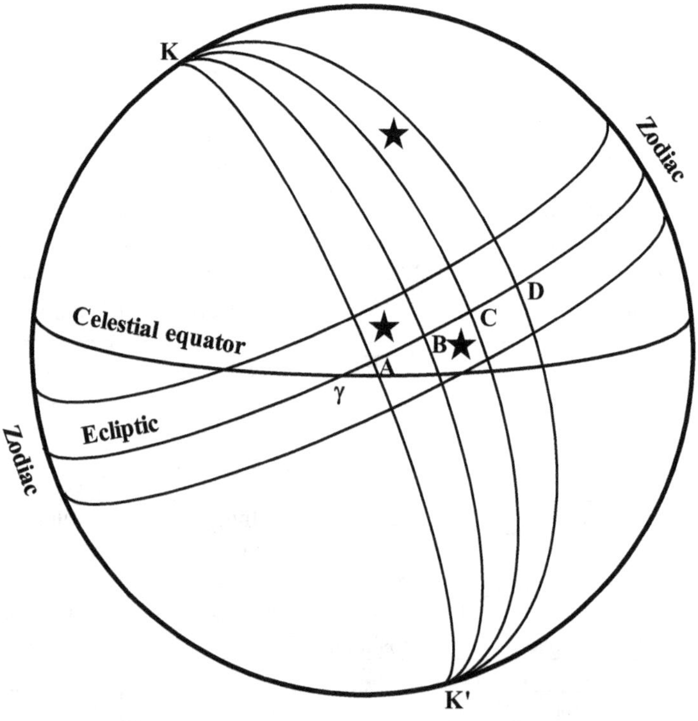

Figure 4.1: The division of celestial sphere in nakṣatra zones

Each nakṣatra zone comprises of the area bound by two great semi-circles passing through K and K′ and through its boundaries on the ecliptic such as KAK′BK, KBK′CK, KCK′DK, and so on. It should be noted that in the ancient Indian system it was not necessary for the yogatārā and stars belonging to a nakṣatra to fall within the zodiac, which is a region spanning 8° on each side of the ecliptic. Figure 4.2 shows the order of nakṣatras in a cyclic manner.

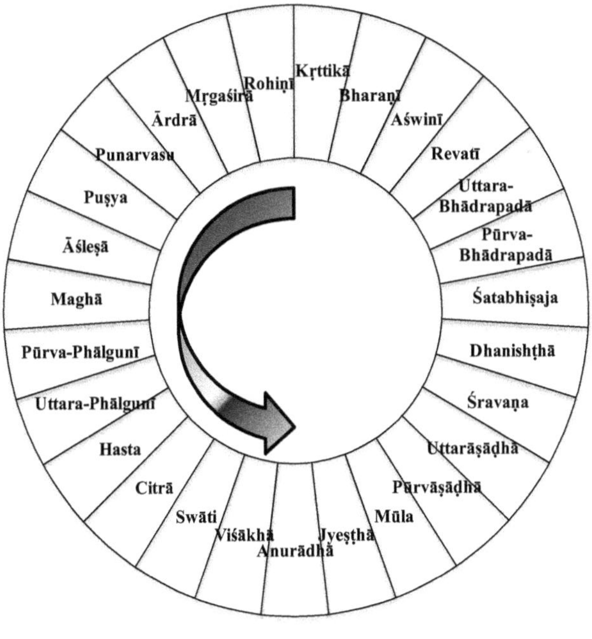

Figure 4.2: The order of 27 nakṣatras

Each nakṣatra was recognized by a group of stars and out of these stars, one star was designated a conjunction star (yogatārā) in Hindu astronomy.

Notes:

1. Whitney (1905): 907-909.
2. Keith (1914): 349.
3. Atharvaveda Saṃhitā 19.7.1-5
4. Taittirīya Saṃhitā iv.4.10
5. Ṛk Vedāṅga Jyotiṣa 25-28, Yajus Vedāṅga Jyotiṣa 32-35.

"Mortal as I am, I know that I am born for a day. But when I follow at my pleasure the serried multitude of the stars in their circular course, my feet no longer touch the earth."

— Ptolemy

5. The Coordinates of Yogatārās

Vedic texts divide the ecliptic in 28 or 27 divisions called nakṣatras. Later astronomical texts such as Sūrya Siddhānta adopted the system of 27 equal divisions. In this system, each nakṣatra has a span of $13^{\circ} 20'$. In addition to the nakṣatra being a geometrical division of ecliptic spanning over a certain segment of ecliptic, each nakṣatra is identified by a star or group of stars. Each nakṣatra is also assigned a yogatārā or junction star out of the group of stars belonging to the particular nakṣatra. If a nakṣatra has only one star in its group, then that star is the yogatārā of that nakṣatra by default. Astronomical text Sūrya Siddhānta gives the coordinates of each yogatārā. Other astronomical texts also give the coordinates of yogatārās. The coordinates given in different texts differ slightly for certain yogatārās and are exactly same for certain other yogatārās.

5.1. Textual information

The coordinates of yogatārās given in Sūrya Siddhānta 8.1-9 are shown in Table 5.1. Three different translations of Sūrya Siddhānta were consulted [1-3]. All translations provide identical information regarding the raw data given in Sūrya Siddhānta and how the data is to be interpreted. The longitude data is given indirectly using a term "Swabhoga". Instead of giving swabhoga values, one tenth of these values are given as shown in the column "Data" in Table 5.1. After multiplying by ten, these values yield swabhoga in arcminutes, which is then converted to degree and arcminutes as shown under the column "Swabhoga" in Table 5.1. Swabhoga represents the relative longitude and is to be added to the longitude of the beginning of the respective nakṣatra to obtain the longitude of the given yogatārā. The list of nakṣatras begins with Aświnī and ends with Revatī. Since Sūrya Siddhānta follows the system of 27 equal divisions of the ecliptic, each nakṣatra has a span of 13° 20′. The beginning and end points of each nakṣatra can then be calculated as shown under the column "Span". Adding relative longitude to the longitude of the beginning point of respective nakṣatra yields the longitude of the yogatārā and is called dhruvaka in Sūrya Siddhānta. The calculated values of dhruvaka are shown in the column "Dhruvaka" in Table 5.1.

Sūrya Siddhānta calls the latitude Vikṣepa and gives the value and direction relative to ecliptic directly as shown in the column "Vikṣepa" in Table 1. Dhruvaka and vikṣepa are universally translated as polar longitude and polar latitude respectively.

Table 5.1: The coordinates of yogatārās [1-3]

No.	Nakṣatra yogatārā	Swabhoga (relative longitude)		
		Data	Arcminutes	Deg.-min.
1	Aświnī	48	480	8° 0′
2	Bharaṇī	40	400	6° 40′
3	Kṛttikā	65	650	10° 50′
4	Rohiṇī	57	570	9° 30′
5	Mṛgaśirā	58	580	9° 40′
6	Ārdrā	4	40	0° 40′
7	Punarvasu	78	780	13° 0′
8	Puṣya	76	760	12° 40′
9	Āśleṣā	14	140	2° 20′
10	Maghā	54	540	9° 0′
11	Pūrva-phālgunī	64	640	10° 40′
12	Uttara-phālgunī	50	500	8° 20′
13	Hasta	60	600	10° 0′
14	Citrā	40	400	6° 40′
15	Swāti	74	740	12° 20′
16	Viśākhā	78	780	13° 0′
17	Anurādhā	64	640	10° 40′
18	Jyeṣṭhā	14	140	2° 20′
19	Mūla	6	60	1° 0′
20	Pūrvāṣāḍhā	4	40	0 ° 40′
21	Uttarāṣāḍhā	Middle of Pūrvāṣāḍhā		
22	Abhijit	End of Pūrvāṣāḍhā		
23	Śravaṇa	End of Uttarāṣāḍhā		
24	Dhaniṣṭhā	Junction of 3rd and 4th quarter of Śravaṇa		
25	Śatabhiṣaja	80	800	13° 20′
26	Pūrva-bhādrapadā	36	360	6° 0′
27	Uttara-bhādrapadā	22	220	3° 40′
28	Revatī	79	790	13° 10′

Table 5.1 (continued)

No.	Nakṣatra yogatārā	Span	Dhruvaka	Vikṣepa
		Deg.-min.	Deg.-min.	Deg.-min.
1	Aświnī	0° 0′ – 13° 20′	8° 0′	10° 0′ N
2	Bharaṇī	13° 20′ – 26°40′	20° 0′	12° 0′ N
3	Kṛttikā	26° 40′ - 40° 0′	37° 30′	5° 0′ N
4	Rohiṇī	40° 0′ - 53° 20′	49° 30′	5° 0′ S
5	Mṛgaśirā	53° 20′ - 66° 40′	63° 0′	10° 0′ S
6	Ārdrā	66° 40′ - 80° 0′	67° 20′	9° 0′ S
7	Punarvasu	80° 0′ - 93° 20′	93° 0′	6° 0′ N
8	Puṣya	93° 20′ - 106° 40′	106° 0′	0° 0′
9	Āśleṣā	106° 40′ - 120° 0′	109° 0′	7° 0′ S
10	Maghā	120° 0′ - 133° 20′	129° 0′	0° 0′
11	Pūrva-phālgunī	133° 20′ - 146° 40′	144° 0′	12° 0′ N
12	Uttara-phālgunī	146° 40′ - 160° 0′	155° 0′	13° 0′ N
13	Hasta	160° 0′ - 173° 20′	170° 0′	11° 0′ S
14	Citrā	173° 20′ - 186° 40′	180° 0′	2° 0′ S
15	Swāti	186° 40′ - 200° 0′	199° 0′	37° 0′ N
16	Viśākhā	200° 0′ - 213° 20′	213° 0′	1°30′ S
17	Anurādhā	213° 20′ - 226° 40′	224° 0′	3° 0′ S
18	Jyeṣṭhā	226° 40′ - 240° 0′	229° 0′	4° 0′ S
19	Mūla	240° 0′ - 253° 20′	241° 0′	9° 0′ S
20	Pūrvāṣāḍhā	253° 20′ - 266° 40′	254° 0′	5° 30′ S
21	Uttarāṣāḍhā	266° 40′ - 280° 0′	260° 0′	5° 0′ S
22	Abhijit	None	266° 40′	60° 0′ N
23	Śravaṇa	280° 0′ - 293° 20′	280° 0′	30° 0′ N
24	Dhaniṣṭhā	293° 20′ - 306° 40′	290° 0′	36° 0′ N
25	Śatabhiṣaja	306° 40′ - 320° 0′	320° 0′	0° 30′ S
26	Pūrva-bhādrapadā	320° 0′ - 333° 20′	326° 0′	24° 0′ N
27	Uttara-bhādrapadā	333° 20′ - 346° 40′	337° 0′	26° 0′ N
28	Revatī	346° 40′ - 360° 0′	359° 50′	0° 0′

After Uttarāṣāḍhā nakṣatra (at number 21 in the list), Abhijit nakṣatra is listed, but no span is given to this nakṣatra. Abhijit nakṣatra was part of 28 nakṣatra system, but was dropped from the list in 27 nakṣatra system, and hence has no span in this system. The longitudes of the yogatārās of Uttarāṣāḍhā, Abhijit, Śravaṇa, and Dhaniṣṭhā nakṣatras are given differently in terms of their positions relative to other nakṣatras as shown in Table 5.1. The yogatārās of Uttarāṣāḍhā and Dhaniṣṭhā fall outside the span of their respective nakṣatras.

Burgess interpreted the coordinates given in Sūrya Siddhānta as polar coordinates, which is universally accepted [4]. These coordinates depend on the position of the North Celestial Pole, which changes over time due to precession. However, the coordinates given in different astronomical texts are nearly same, even though the texts were written many centuries apart. The change in coordinates would have been obvious to the astronomers, if they had the skill to measure the coordinates of the stars. This has resulted in the opinion that Indian astronomers were borrowers from the west and incapable of making accurate astronomical observations. Pingree and Morrissey write the following about the Indian astronomers:

> We hope thereby to demonstrate three things: that there is no basis for identifying the stars included in the Vedic nakṣatras, and therefore no grounds for comparing them, for example, with the Chinese lunar mansions; that the catalogue of stars found in Paitāmahasiddhānta, which is almost exclusively the basis of the rest of the Indian tradition, since it is at the beginning of the Indian

> attempts to provide coordinates and uses a coordinate system derived from Greek astronomy, is more likely to be an Indian adaptation of a Greek star catalogue than to be based on observations that were made in India; and that the ineptitude with which Indians historically tried to 'correct' these coordinates militates against any theory that is founded upon the idea that the Indians of medieval period were experts in astronomical observation. ... Our apparent success in finding "identifications" for Lalla's star catalogue, wherein the coordinates are so clearly a mixture of ecliptic and polar values, shows the futility of attaching any credence to them. ... Whichever stars the author of Sūryasiddhānta intended to indicate, he was incapable of determining their coordinates accurately, ... It is most astonishing to see an astronomer convert λ^* into λ and call the latter λ^*; even more astonishing is to see him take λ to be λ^*, convert it on that assumption into another λ, and to assert that this wrongly derived λ is λ^*! There is no excuse for Āryabhaṭa's coordinates. ... The impression of incompetence does not disappear when we examine our last star catalogue, that which Gaṇeṣa incorporated into his Grahalāghava (XI 1-5) in 1520. ... Therefore, either Gaṇeṣa was also incompetent, or he intended to give the coordinates of a different set of stars. ... We must conclude from this survey that the Indians did not observe the positions of the stars with accuracy; by implication, they also did not observe those of the planets with accuracy. [5]

It is obvious that the interpretation of coordinates given in Indian astronomical texts as polar coordinates has resulted in Indian astronomers being called incompetent.

In my recently published peer-reviewed paper I have proposed that the coordinates of yogatārās given in the Indian astronomical texts are ecliptic coordinates and since these coordinates don't change appreciably over time in sidereal ecliptic coordinate system, Indian astronomers relied on the coordinates received from earlier astronomers [6]. It is for this reason that the coordinates of yogatārās were not updated. To better appreciate this point of view, it is important to understand how the coordinates of stars are measured.

Notes

1. Burgess (1860)
2. Śrīvāstava (1982)
3. Siṃha (1986)
4. Burgess (1860)
5. Pingree and Morrissey (1989)
6. Roy (2019)

"Just like a GPS, the universe sends you signs to show you the best course. If you follow the flow, you get where you want with ease and happiness. If you miss a turn, the road becomes longer and harder."

— Charbel Tadros

6. GPS of the Stars

The specification of the position of a star is similar to the specification of a point on earth. Any point on earth can be specified by a pair of spherical coordinates called latitude and longitude. Since mathematically any point on a sphere is identical to another point on the sphere, latitude and longitude can only be specified by making two specific choices, one for latitude and another for longitude. In spherical geometry, these choices are named fundamental plane for measuring latitude and zero point for measuring longitude.

The fundamental plane is a plane of reference that divides the sphere into two hemispheres. The latitude of a point is the angle between the fundamental plane and the line joining the point to the centre of sphere. For the Global Positioning System (GPS), the fundamental plane is the equator. The latitude of any point on the equator is zero. The latitude of North Pole is 90° and the latitude of South

Pole is -90°. The latitude of all points on earth falls between 90° and -90°.

The zero point is a point on the fundamental plane from which the longitude is measured along the fundamental plane. A meridian connects the points of equal longitude between the North Pole and the South Pole of the fundamental plane. For the Global Positioning System (GPS), the zero point is the intersection of equator and prime meridian, which is the meridian passing through Greenwich. The longitude is defined to be 0° at prime meridian.

Depending on the choice of the fundamental plane and the zero point, there are many celestial coordinate systems for specifying the position of a star. The most widely used celestial coordinate systems are Horizontal, Equatorial, and Ecliptic coordinate systems.

6.1 Horizontal coordinates

Just as the coordinates of a point on earth are measured assuming the earth to be a perfect sphere, the coordinates of a star are measured assuming the star to be located on a perfect sphere called celestial sphere of arbitrarily large radius. Just like the GPS, the coordinates are measured in angles, not distance.

Horizontal coordinates are specified by providing altitude and azimuth. Figure 6.1 illustrates the horizontal coordinate system. In this picture, O is the location of the observer, ON is the north direction, OE is the east direction, OS is the south direction, OW is the west direction, Z is the zenith, S′ is the location of a star, and Z′ is the nadir. Z is the point on celestial sphere directly overhead the observer and Z′ is the point on celestial sphere directly underneath the observer. The great circle connecting N, E, W, and S is the horizon. The vertical circle passing through N, Z, and S is called the meridian. Altitude of the star is given by S′A and the azimuth of the star is given by NA. In horizontal coordinate system, the altitude is determined by measuring the angular distance from the horizon to the star along the great circle passing through the star and the zenith. Azimuth is determined by measuring the angular distance along the horizon from the north direction going eastward.

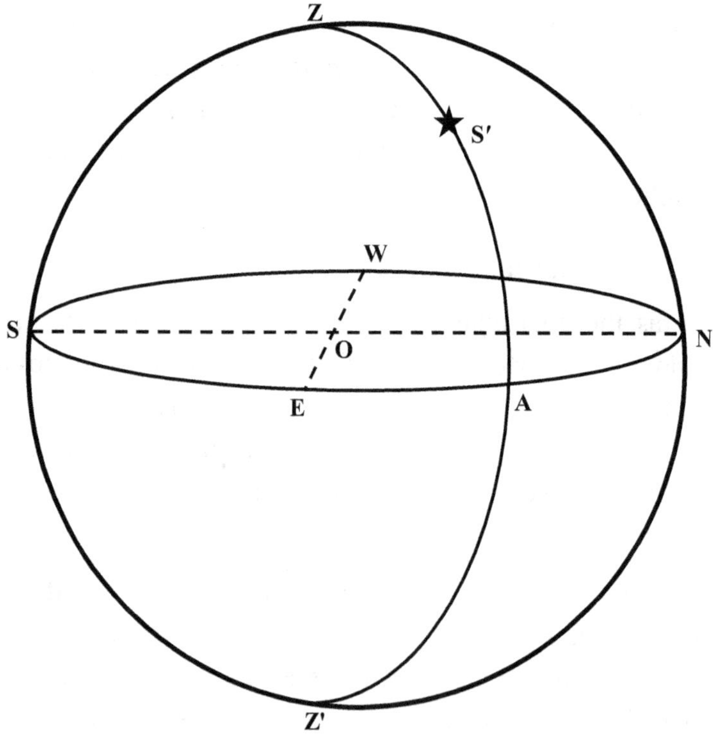

Figure 6.1: Illustration of horizontal coordinate system

In the horizontal coordinate system, the fundamental plane is the observer's horizon and the zero point is observer's north point on the horizon, point N in Figure 6.1.

6.2 Equatorial coordinates

Equatorial coordinate system measures the coordinates from and along the celestial equator, which is the projection of the earth's equator on the celestial sphere. The poles of the celestial equator are called North Celestial Pole (NCP) and South Celestial Pole (SCP), which are the projections of the North Pole and South Pole on the celestial sphere respectively.

Equatorial coordinates are specified by providing declination and right ascension. Figure 6.2 illustrates the equatorial coordinate system. In this picture P is the North Celestial Pole (NCP), ♈ is the first point of Aries, S is the location of a star, and P′ is the South Celestial Pole (SCP). Declination of the star is given by SA and the right ascension of the star is given by ♈A. In equatorial coordinate system, the declination is determined by measuring the angular distance from the celestial equator to the star along the great circle passing through the star and the North Celestial Pole. Right ascension is determined by measuring the angular distance along the celestial equator from the first point of Aries to the intersection of the celestial equator and the great circle passing through the star and the North Celestial Pole.

In the equatorial coordinate system, the fundamental plane is the celestial equator and the zero point is first point of Aries, point ♈ in Figure 6.2.

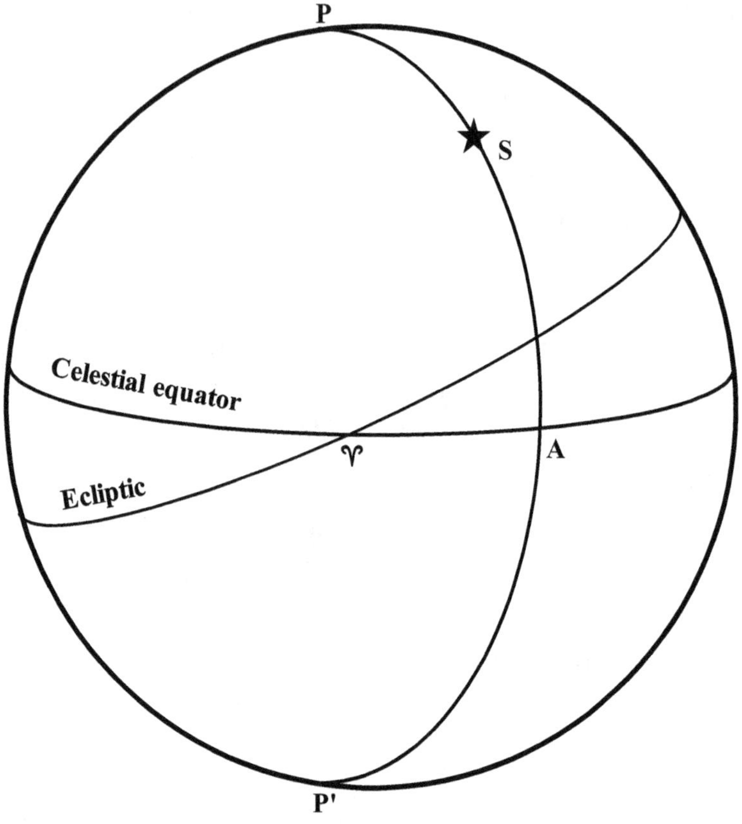

Figure 6.2: Illustration of equatorial coordinate system

6.3 Ecliptic coordinates

Ecliptic coordinate system measures the coordinates from and along the ecliptic, which is a great circle on the celestial sphere representing the sun's apparent path during a year. The poles of the ecliptic are called North Ecliptic Pole (NEP) and South Ecliptic Pole (SEP).

Ecliptic coordinates are specified by providing ecliptic latitude and ecliptic longitude. Figure 6.3 illustrates the ecliptic coordinate system. In this picture K is the North Ecliptic Pole (NEP), ♈ is the first point of Aries, S is the location of a star, and K′ is the South Ecliptic Pole (SEP). Ecliptic latitude of the star is given by SA and the ecliptic longitude of the star is given by ♈A.

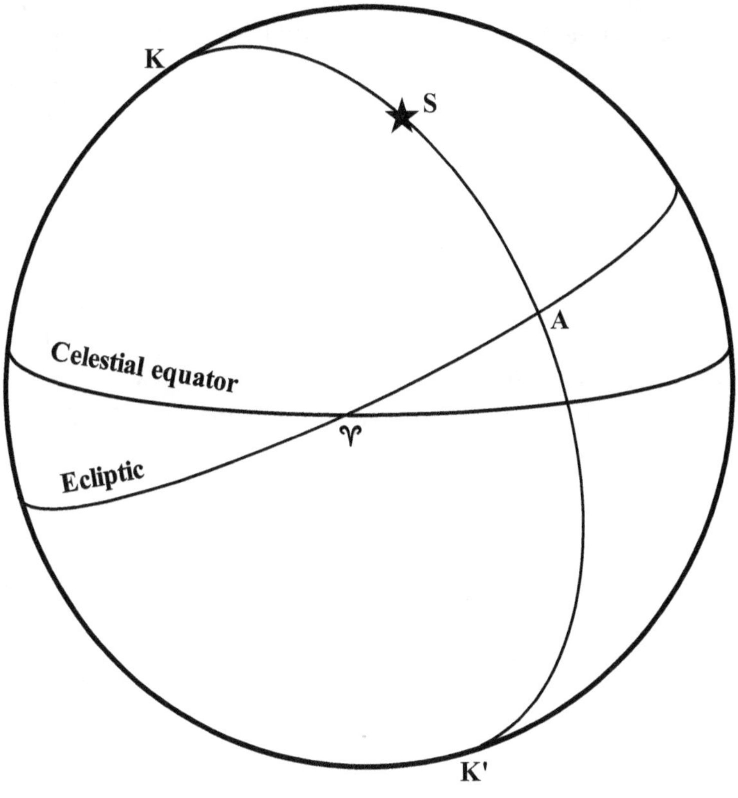

Figure 6.3: Illustration of ecliptic coordinate system

In ecliptic coordinate system, ecliptic latitude is determined by measuring the angular distance from ecliptic to the star along the great circle passing through the star and the North Ecliptic Pole. Ecliptic longitude is determined by measuring the angular distance along the ecliptic from the first point of Aries to the intersection of the ecliptic and the great circle passing through the star and the North Ecliptic Pole.

In the ecliptic coordinate system, the fundamental plane is the ecliptic and the zero point is first point of Aries, point ♈ in Figure 6.3.

6.4 Polar coordinates

It is currently accepted that the coordinates of yogatārās given in Sūrya Siddhānta are polar longitudes and latitudes. The terms polar longitude and polar latitude were coined by Burgess in his translation of Sūrya Siddhānta [1], which uses the term Dhruvaka for longitude and Vikṣepa for latitude. Burgess has identified Dhruvaka and Vikṣepa as polar longitude and polar latitude respectively. The concept of polar coordinates of stars as illustrated by Burgess is shown in Figure 6.4. To determine the polar longitude and latitude of a star (S or S′), a segment of circle of declination (PSca or Pc′a′S) is drawn from North Celestial Pole (P) passing through the star up to the ecliptic. Polar latitude is the angular distance of the star (Sa or S′a′) from the ecliptic along the circle of declination. Polar longitude is the angular distance (La or La′) from reference point (L) on the ecliptic and the point of intersection of the ecliptic with the circle of declination passing through the star (a or a′).

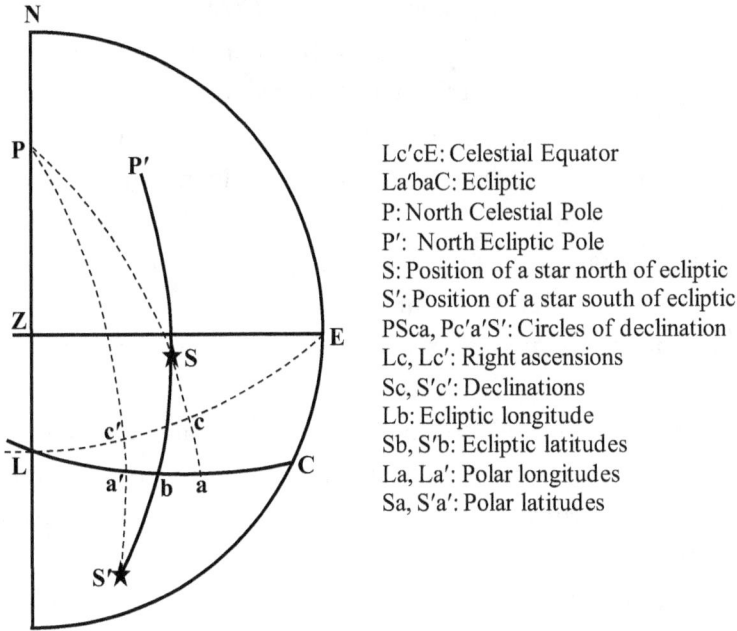

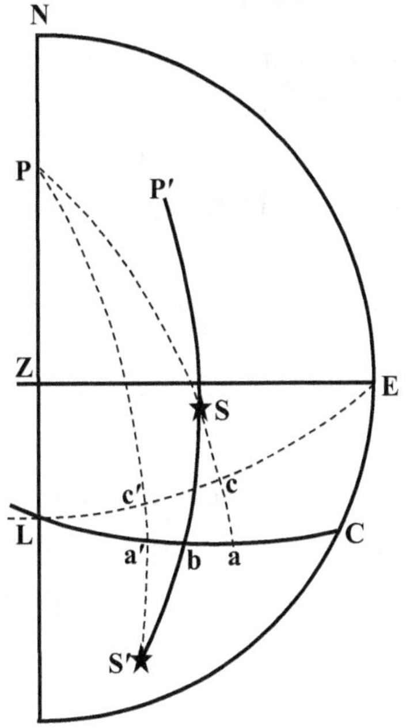

Lc′cE: Celestial Equator
La′baC: Ecliptic
P: North Celestial Pole
P′: North Ecliptic Pole
S: Position of a star north of ecliptic
S′: Position of a star south of ecliptic
PSca, Pc′a′S′: Circles of declination
Lc, Lc′: Right ascensions
Sc, S′c′: Declinations
Lb: Ecliptic longitude
Sb, S′b: Ecliptic latitudes
La, La′: Polar longitudes
Sa, S′a′: Polar latitudes

Figure 6.4: Illustration of polar longitude and latitude of stars by Burgess [1]

It should be noted that this whole geometrical construction for determining polar longitude and latitude is very artificial. In ecliptic coordinate system, ecliptic latitudes are determined by measuring angular distances from ecliptic along the great circle passing through the North Ecliptic Pole. In celestial coordinate system, declinations are determined by measuring angular distances from celestial equator along the great circle passing through the North Celestial Pole. In every coordinate system, the latitude is measured respective to the corresponding pole. In the artificial construct of polar latitude, the angular distance is measured from the ecliptic along the great circle that does not pass through the pole of ecliptic (North Ecliptic Pole), but passes through North Celestial Pole instead. Burgess has justified this artificial construction by taking the meaning of Dhruvaka as pertaining to Dhruva or pole star, and therefore he has drawn great circle passing through North Celestial Pole. In accordance with Dhruvaka, Burgess has postulated that Viksepa means polar latitude. There is absolutely nothing in any astronomical text that describes this method of measuring longitude and latitude. The term viksepa has been used many times in Sūrya Siddhānta such as in 2.6, 2.63 and 7.7, and in all these places viksepa has not been interpreted as polar latitude even by Burgess [1]. Moreover, Dhruva also means fixed or not moving and thus Dhruvaka simply means fixed longitude. Many Indian astronomers such as Bhāskara in Mahābhāskarīya refer to the coordinates of yogatārās explicitly as ecliptic longitude and latitude. Pingree and Morrissey write:

> What is remarkable about Bhāskara's ecliptic coordinates is that, in most cases, they are within 1° of the Paitāmaha's polar coordinates; this is the case for the longitudes of nos. 1, 4, 5, 8, 10, 12, 16, 18, 19, and 28, and for the latitudes of nos. 1, 2, 3, 4, 5, 6, 7, 8, 9, 10, 11, 12, 15, 16, 17, 18, 19, 23, 24, 25, 26, 27, and 28 – that is, for 33 out of 54 possibilities. This fact denies to Bhāskara the possibility of his having himself made independent observations, or of his having used a source based on independent observations. This lack of observational input is emphasized by the fact that his changes of the Paitāmaha's coordinates lead either to worse results or to dimmer stars or to both. [2]

It is surprising that despite the emphatic declaration by Indian astronomers that they were using ecliptic coordinates, no one has challenged the prevailing view that Indian astronomers were using polar coordinates. First, the information provided by Indian astronomers has been incorrectly interpreted and then that false interpretation has been used to claim that Indian astronomers were inept and did not know how to measure the positions of stars and planets. If that is the case, then correct framework needs to be developed in which the coordinates given by Indian astronomers make better sense and currently accepted identifications of yogatārās need to be reassessed to check if other stars fit the given coordinates better.

Notes

1. Burgess (1860)
2. Pingree and Morrissey (1989)

"The cosmos is within us. We are made of star-stuff. We are a way for the universe to know itself."
— Carl Sagan

7. Current Identification of Nakṣatra Stars

Based on his assumption that coordinates given in Sūrya Siddhānta are polar longitudes and latitudes, Burgess [1] identified the yogatārās as shown in Table 7.1. These identifications are currently accepted by most scholars. Ecliptic coordinates (J2000.0) of these yogatārās are also given in Table 7.1. The data for ecliptic coordinates (J2000.0) were obtained using Stellarium software by setting the date to January 1, 2000 at 12:00 noon and noting the ecliptic longitudes and latitudes by selecting the specific stars.

Some astronomical texts give the number of stars in each nakṣatra. Table 7.2 shows the number of stars in each nakṣatra according to Nakṣatrakalpa of Atharvaveda, Śārdūlakarṇāvadāna, and Gargasaṃhitā as compiled by Pingree and Morrissey [2].

The stars belonging to these nakṣatra star groups have been listed by Kaye [3]. Out of these star groups, yogatārās have also been specified by Kaye, which match exactly with the yogatārās identified by Burgess [1].

Table 7.1: Identification of yogatārās by Burgess [1]

No.	Nakṣatra	Yogatārā*	Ecliptic Longitude**	Ecliptic Latitude**
1	Aświnī	β Ari	33° 58′	8° 29′ N
2	Bharaṇī	35 Ari	46° 56′	11° 19′ N
3	Kṛttikā	η Tau	60° 00′	4° 03′ N
4	Rohiṇī	α Tau	69° 47′	5° 28′ S
5	Mṛgaśirā	λ Ori	83° 42′	13° 22′ S
6	Ārdrā	α Ori	88° 45′	16° 02′ S
7	Punarvasu	β Gem	113° 13′	6° 41′ N
8	Puṣya	δ Cnc	128° 43′	0° 05′ N
9	Āśleṣā	ε Hya	132° 21′	11° 06′ S
10	Maghā	α Leo	149° 50′	0° 28′ N
11	Pūrva-phālgunī	δ Leo	161° 19′	14° 20′ N
12	Uttara-phālgunī	β Leo	171° 37′	12° 16′ N
13	Hasta	δ Crv	193° 27′	12° 12′ S
14	Citrā	α Vir	203° 50′	2° 03′ S
15	Swāti	α Boo	204° 14′	30° 44′ N
16	Viśākhā	ι Lib	231° 00′	1°51′ S
17	Anurādhā	δ Sco	242° 34′	1° 59′ S
18	Jyeṣṭhā	α Sco	249° 46′	4° 34′ S
19	Mūla	λ Sco	264° 35′	13° 47′ S
20	Pūrvāṣāḍhā	δ Sgr	274° 35′	6° 28′ S
21	Uttarāṣāḍhā	σ Sgr	282° 23′	3° 27′ S
22	Abhijit	α Lyr	285° 19′	61° 44′ N
23	Śravaṇa	α Aql	301° 47′	29° 18′ N
24	Dhaniṣṭhā	β Del	316° 20′	31° 55′ N
25	Śatabhiṣaja	λ Aqr	341° 35′	0° 23′ S
26	Pūrva-bhādrapadā	α Peg	353° 29′	19° 24′ N
27a	Uttara-bhādrapadā***	α And	14° 19′	25° 41′ N
27b	Uttara-bhādrapadā***	γ Peg	9° 09′	12° 36′ N
28	Revatī	ζ Psc	19° 53′	0° 13′ S

* As identified by Burgess [1]

** J2000.0 ecliptic coordinates based on Stellarium software.

*** For Uttara-bhādrapadā, longitude matches γ Pegasi, while latitude matches α Andromeda.

Table 7.2: Number of stars in nakṣatras [2]

<table>
<tr><th></th><th>Nakṣatra</th><th>Nakṣatra-kalpa</th><th>Śārdūla-karṇāvadāna</th><th>Garga-saṃhitā</th></tr>
<tr><td>1</td><td>Aśvinī</td><td>2</td><td>2</td><td>2</td></tr>
<tr><td>2</td><td>Bharaṇī</td><td>3</td><td>3</td><td>3</td></tr>
<tr><td>3</td><td>Kṛttikā</td><td>6</td><td>6</td><td>6</td></tr>
<tr><td>4</td><td>Rohiṇī</td><td>1</td><td>5</td><td>5</td></tr>
<tr><td>5</td><td>Mṛgaśirā</td><td>3</td><td>3</td><td>3</td></tr>
<tr><td>6</td><td>Ārdrā</td><td>1</td><td>1</td><td>1</td></tr>
<tr><td>7</td><td>Punarvasu</td><td>2</td><td>2</td><td>2</td></tr>
<tr><td>8</td><td>Puṣya</td><td>1</td><td>3</td><td>1</td></tr>
<tr><td>9</td><td>Āśleṣā</td><td>6</td><td>1</td><td>6</td></tr>
<tr><td>10</td><td>Maghā</td><td>6</td><td>5</td><td>6</td></tr>
<tr><td>11</td><td>Pūrva-Phālgunī</td><td rowspan="2">4</td><td>2</td><td>2</td></tr>
<tr><td>12</td><td>Uttara-Phālgunī</td><td>2</td><td>2</td></tr>
<tr><td>13</td><td>Hasta</td><td>5</td><td>5</td><td>5</td></tr>
<tr><td>14</td><td>Citrā</td><td>1</td><td>1</td><td>1</td></tr>
<tr><td>15</td><td>Svāti</td><td>1</td><td>1</td><td>1</td></tr>
<tr><td>16</td><td>Viśākhā</td><td>2</td><td>2</td><td>2</td></tr>
<tr><td>17</td><td>Anurādhā</td><td>4</td><td>4</td><td>4</td></tr>
<tr><td>18</td><td>Jyeṣṭhā</td><td>1</td><td>3</td><td>3</td></tr>
<tr><td>19</td><td>Mūla</td><td>7</td><td>7</td><td>6</td></tr>
<tr><td>20</td><td>Pūrvāṣāḍhā</td><td>8</td><td>4</td><td>4</td></tr>
<tr><td>21</td><td>Uttarāṣāḍhā</td><td></td><td>4</td><td>4</td></tr>
<tr><td>22</td><td>Abhijit</td><td>1</td><td>3</td><td>3</td></tr>
<tr><td>23</td><td>Śravaṇa</td><td>3</td><td>3</td><td>3</td></tr>
<tr><td>24</td><td>Dhaniṣṭhā</td><td>5</td><td>4</td><td>4</td></tr>
<tr><td>25</td><td>Śatabhiṣaja</td><td>1</td><td>1</td><td>1</td></tr>
<tr><td>26</td><td>Pūrva-Bhādrapadā</td><td rowspan="2">4</td><td>2</td><td>2</td></tr>
<tr><td>27</td><td>Uttara-Bhādrapadā</td><td>2</td><td>2</td></tr>
<tr><td>28</td><td>Revatī</td><td>1</td><td>1</td><td>4</td></tr>
</table>

The nakṣatra star groups identified by Kaye are shown in Figures 7.1 to 7.11. Proper name or HIP, Bayer designation, Flamsteed designation, apparent magnitude, ecliptic longitude (J2000.0), and ecliptic latitude (J2000.0) of these stars are shown in Tables 7.3 to 7.30. Yogatārās are shown in bold letters and numbers in each table. The data for J2000.0 ecliptic longitudes and latitudes were obtained using Stellarium software by setting the date to January 1, 2000 at 12:00 noon and noting the ecliptic longitudes and latitudes by selecting the specific stars. In each table, stars are numbered in the order of increasing longitude.

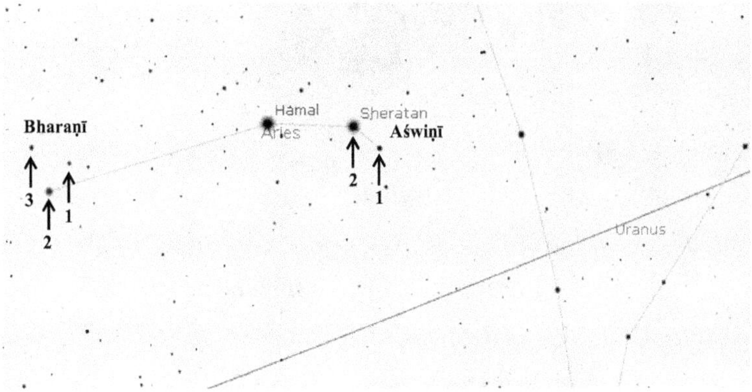

Figure 7.1: Aświnī, and Bharaṇī nakṣatras

Table 7.3: Aświnī nakṣatra

	Proper name	Bayer designation	Flamsteed designation	Apparent Magnitude	Ecliptic longitude	Ecliptic latitude
1.	Mesarthim	γ1 Ari	5 Ari	4.50	33° 11′	7° 10′
2.	**Sheratan**	**β Ari**	**6 Ari**	**2.60**	**33° 58′**	**8° 29′**

Table 7.4: Bharaṇī nakṣatra

	Proper name	Bayer designation	Flamsteed designation	Apparent Magnitude	Ecliptic longitude	Ecliptic latitude
1.	**Barani II**		**35 Ari**	**4.65**	**46° 56′**	**11° 19′**
2.	Bharani		41 Ari	3.60	48° 12′	10° 27′
3.	Barani III		39 Ari	4.50	48° 22′	12° 29′

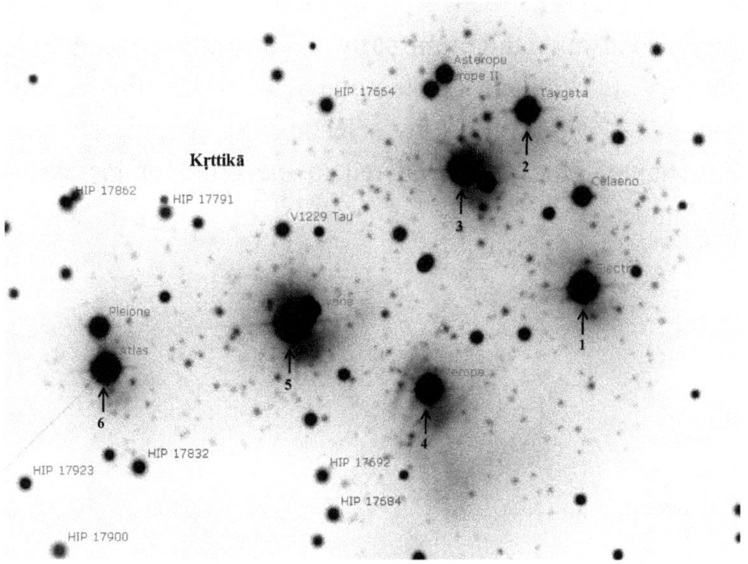

Figure 7.2: Kṛttikā nakṣatra

Table 7.5: Kṛttikā nakṣatra

	Proper name	Bayer designation	Flamsteed designation	Apparent Magnitude	Ecliptic longitude	Ecliptic latitude
1.	Electra		17 Tau	3.70	59° 25′	4° 11′
2.	Taygeta	q Tau	19 Tau	4.30	59° 34′	4° 31′
3.	Maia		20 Tau	3.85	59° 41′	4° 23′
4.	Merope		23 Tau	4.10	59° 42′	3° 57′
5.	**Alcyone**	**η Tau**	**25 Tau**	**2.85**	**60° 00′**	**4° 03′**
6.	Atlas		27 Tau	3.60	60° 21′	3° 55′

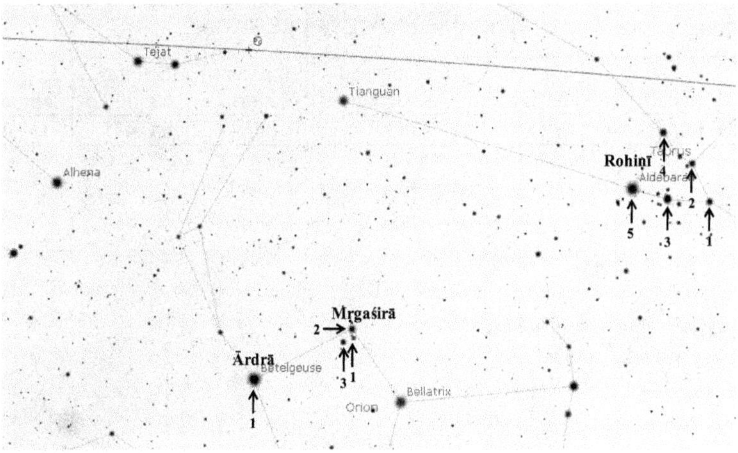

Figure 7.3: Rohiṇī, Mṛgaśirā, and Ārdrā nakṣatras

Table 7.6: Rohiṇī nakṣatra

	Proper name	Bayer designation	Flamsteed designation	Apparent Magnitude	Ecliptic longitude	Ecliptic latitude
1.	Hyadum I	γ Tau	54 Tau	3.65	65° 48′	-5° 44′
2.	Hyadum II	δ1 Tau	61 Tau	3.75	66° 52′	-3° 58′
3.	Hyadum IV	θ2 Tau	78 Tau	3.40	67° 58′	-5° 50′
4.	Ain	ε Tau	74 Tau	3.50	68° 28′	-2° 34′
5.	**Aldebaran**	**α Tau**	**87 Tau**	**0.85**	**69° 47′**	**-5° 28′**

Table 7.7: Mṛgaśirā nakṣatra

	Proper name/HIP	Bayer designation	Flamsteed designation	Apparent Magnitude	Ecliptic longitude	Ecliptic latitude
1.	Heka	ϕ1 Ori	37 Ori	4.35	83° 36′	-13° 49′
2.	**Meissa**	**λ Ori**	**39 Ori**	**3.50**	**83° 42′**	**-13° 22′**
3.	HIP 26366	ϕ2 Ori	40 Ori	4.05	84° 08′	-14° 02′

Table 7.8: Ārdrā nakṣatra

	Proper name	Bayer designation	Flamsteed designation	Apparent Magnitude	Ecliptic longitude	Ecliptic latitude
1.	**Betelgeuse**	**α Ori**	**58 Ori**	**0.45**	**88° 45′**	**-16° 02′**

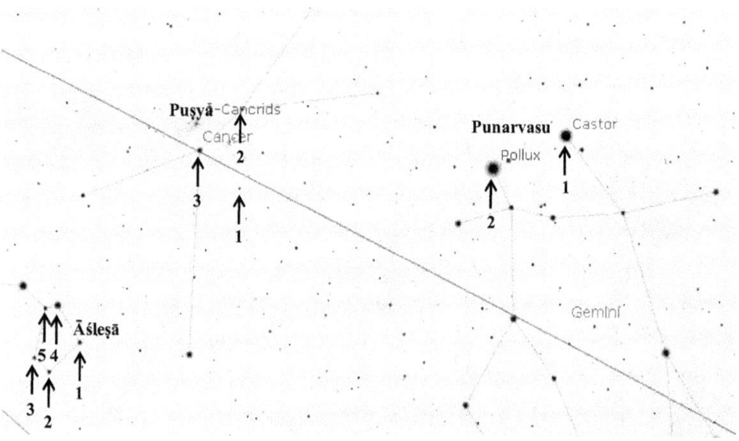

Figure 7.4: Punarvasu, Puṣya, and Āśleṣā nakṣatras

Table 7.9: Punarvasu nakṣatra

	Proper name	Bayer designation	Flamsteed designation	Apparent Magnitude	Ecliptic longitude	Ecliptic latitude
1.	Castor	α Gem	66 Gem	1.90	110° 14′	10° 06′
2.	**Pollux**	**β Gem**	**78 Gem**	**1.15**	**113° 13′**	**6° 41′**

Table 7.10: Puṣya nakṣatra

	Proper name/HIP	Bayer designation	Flamsteed designation	Apparent Magnitude	Ecliptic longitude	Ecliptic latitude
1.	HIP 41822	θ Cnc	31 Cnc	5.30	125° 44′	-0° 46′
2.	Asellus Borealis	γ Cnc	43 Cnc	4.65	127° 32′	3° 11′
3.	**Asellus Australis**	**δ Cnc**	**47 Cnc**	**3.90**	**128° 43′**	**0° 05′**

Table 7.11: Āśleṣā nakṣatra

	Proper name/HIP	Bayer designation	Flamsteed designation	Apparent Magnitude	Ecliptic longitude	Ecliptic latitude
1.	Minazal I	δ Hya	4 Hya	4.10	130° 18′	-12° 24′
2.	Minkalshuja	σ Hya	5 Hya	4.45	131° 13′	-14° 36′
3.	Minazal II	η Hya	7 Hya	4.30	132° 18′	-14° 15′
4.	**Minazal III**	**ε Hya**	**11 Hya**	**3.40**	**132° 21′**	**-11° 06′**
5.	Minazal IV	ρ Hya	13 Hya	4.35	132° 55′	-11° 33′

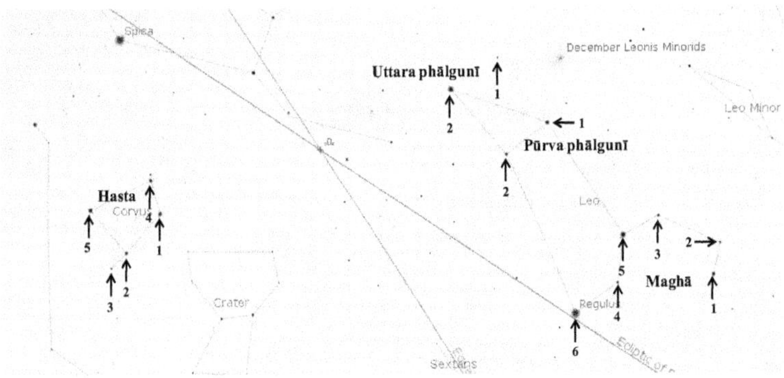

Figure 7.5: Maghā, Pūrva-phālgunī, Uttara-phālgunī, and Hasta nakṣatras

Table 7.12: Maghā nakṣatra

	Proper name	Bayer designation	Flamsteed designation	Apparent Magnitude	Ecliptic longitude	Ecliptic latitude
1.	Algenubi	ε Leo	17 Leo	2.95	140° 42′	9° 43′
2.	Rasalas	μ Leo	24 Leo	3.85	141° 26′	12° 21′
3.	Adhafera	ζ Leo	36 Leo	3.40	147° 34′	11° 52′
4.	Al Jabhah	η Leo	30 Leo	3.45	147° 54′	4° 52′
5.	Algieba	γ1 Leo	41 Leo	2.20	149° 37′	8° 49′
6.	**Regulus**	**α Leo**	**32 Leo**	**1.35**	**149° 50′**	**0° 28′**

Table 7.13: Pūrva-phālgunī nakṣatra

	Proper name	Bayer designation	Flamsteed designation	Apparent Magnitude	Ecliptic longitude	Ecliptic latitude
1.	**Zosma**	**δ Leo**	**68 Leo**	**2.55**	**161° 19′**	**14° 20′**
2.	Chertan	θ Leo	70 Leo	3.30	163° 25′	9° 40′

Table 7.14: Uttara-phālgunī nakṣatra

	Proper name/HIP	Bayer designation	Flamsteed designation	Apparent Magnitude	Ecliptic longitude	Ecliptic latitude
1.	HIP 57565		93 Leo	4.50	168° 58′	17° 19′
2.	**Denebola**	**β Leo**	**94 Leo**	**2.10**	**171° 37′**	**12° 16′**

Table 7.15: Hasta nakṣatra

	Proper name	Bayer designation	Flamsteed designation	Apparent Magnitude	Ecliptic longitude	Ecliptic latitude
1.	Gienah	γ Crv	4 Crv	2.55	190° 44′	-14° 30′
2.	Minkar	ε Crv	2 Crv	3.00	191° 40′	-19° 40′
3.	Alchiba	α Crv	1 Crv	4.00	192° 15′	-21° 45′
4.	**Algorab**	**δ Crv**	**7 Crv**	**2.90**	**193° 27′**	**-12° 12′**
5.	Kraz	β Crv	9 Crv	2.65	197° 22′	-18° 03′

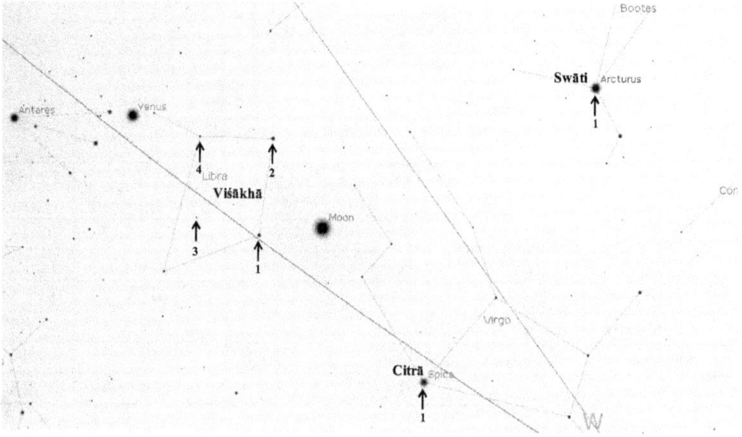

Figure 7.6: Citrā, Swāti and Viśākhā nakṣatras

Table 7.16: Citrā nakṣatra

	Proper name	Bayer designation	Flamsteed designation	Apparent Magnitude	Ecliptic longitude	Ecliptic latitude
1.	**Spica**	**α Vir**	**67 Vir**	**0.95**	**203° 50′**	**-2° 03′**

Table 7.17: Swāti nakṣatra

	Proper name	Bayer designation	Flamsteed designation	Apparent Magnitude	Ecliptic longitude	Ecliptic latitude
1.	**Arcturus**	**α Boo**	**16 Boo**	**0.15**	**204° 14′**	**30° 43′**

Table 7.18: Viśākhā nakṣatra

	Proper name/HIP	Bayer designation	Flamsteed designation	Apparent Magnitude	Ecliptic longitude	Ecliptic latitude
1.	Zuben-elgenubi II	α2 Lib	9 Lib	2.75	225° 05′	0° 20′
2.	Zuben-eschamali	β Lib	27 Lib	2.60	229° 22′	8° 30′
3.	**HIP 74392**	**ι1 Lib**	**24 Lib**	**4.50**	**231° 00′**	**-1° 51′**
4.	Zuben Elakrab	γ Lib	38 Lib	3.90	235° 08′	4° 23′

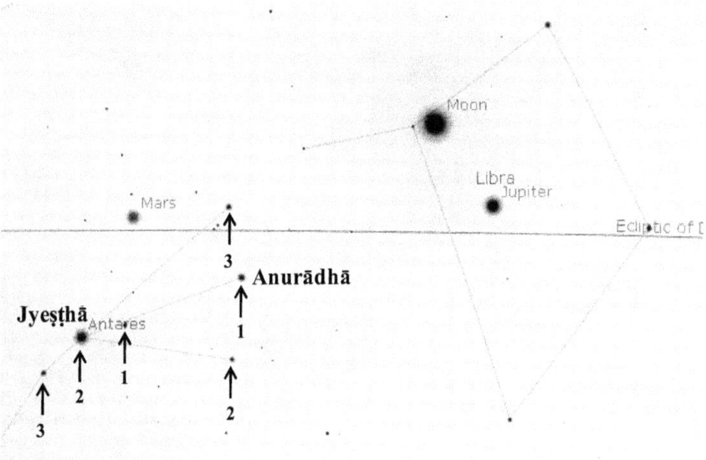

Figure 7.7: Anurādhā, and Jyeṣṭhā nakṣatras

Table 7.19: Anurādhā nakṣatra

	Proper name	Bayer designation	Flamsteed designation	Apparent Magnitude	Ecliptic longitude	Ecliptic latitude
1.	**Dschubba**	**δ Sco**	**7 Sco**	**2.35**	**242° 34′**	**-1° 59′**
2.	Fang	π Sco	6 Sco	2.85	242° 56′	-5° 29′
3.	Acrab	β1 Sco	8 Sco	2.60	243° 11′	1° 00′

Table 7.20: Jyeṣṭhā nakṣatra

	Proper name	Bayer designation	Flamsteed designation	Apparent Magnitude	Ecliptic longitude	Ecliptic latitude
1.	Alniyat	σ Sco	20 Sco	3.05	247° 48′	-4° 02′
2.	**Antares**	**α Sco**	**21 Sco**	**1.05**	**249° 46′**	**-4° 34′**
3.	Alniyat II	τ Sco	23 Sco	2.80	251° 27′	-6° 07′

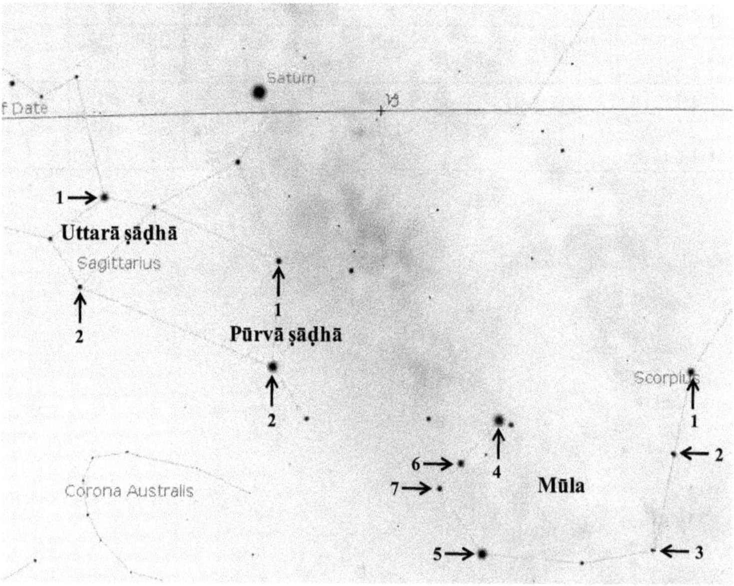

Figure 7.8: Mūla, Pūrvāṣāḍhā, and Uttarāṣāḍhā nakṣatras

Table 7.21: Mūla nakṣatra

	Proper name/HIP	Bayer designation	Flamsteed designation	Apparent Magnitude	Ecliptic longitude	Ecliptic latitude
1.	HIP 82396	ε Sco	26 Sco	2.25	255° 20′	-11° 44′
2.	Tali al Shaulah I	μ1 Sco		3.00	256° 09′	-15° 25′
3.	HIP 82729	ζ2 Sco		3.60	257° 14′	-19° 39′
4.	**Shaula**	**λ Sco**	**35 Sco**	**1.60**	**264° 35′**	**-13° 47′**
5.	Sargas	θ Sco		1.85	265° 36′	-19° 39′
6.	Mula	κ Sco		2.35	266° 28′	-15° 39′
7.	Girtab	ι1 Sco		2.95	267° 31′	-16° 43′

Table 7.22: Pūrvāṣāḍhā nakṣatra

	Proper name/HIP	Bayer designation	Flamsteed designation	Apparent Magnitude	Ecliptic longitude	Ecliptic latitude
1.	**Kaus Media**	**δ Sgr**	**19 Sgr**	**2.70**	**274° 35′**	**-6° 28′**
2.	Kaus Australis	ε Sgr	20 Sgr	1.75	275° 05′	-11° 03′

Table 7.23: Uttarāṣāḍhā nakṣatra

	Proper name/HIP	Bayer designation	Flamsteed designation	Apparent Magnitude	Ecliptic longitude	Ecliptic latitude
1.	**Nunki**	**σ Sgr**	**34 Sgr**	**2.05**	**282° 23′**	**-3° 27′**
2.	Ascella	ζ Sgr	38 Sgr	3.25	283° 38′	-7° 11′

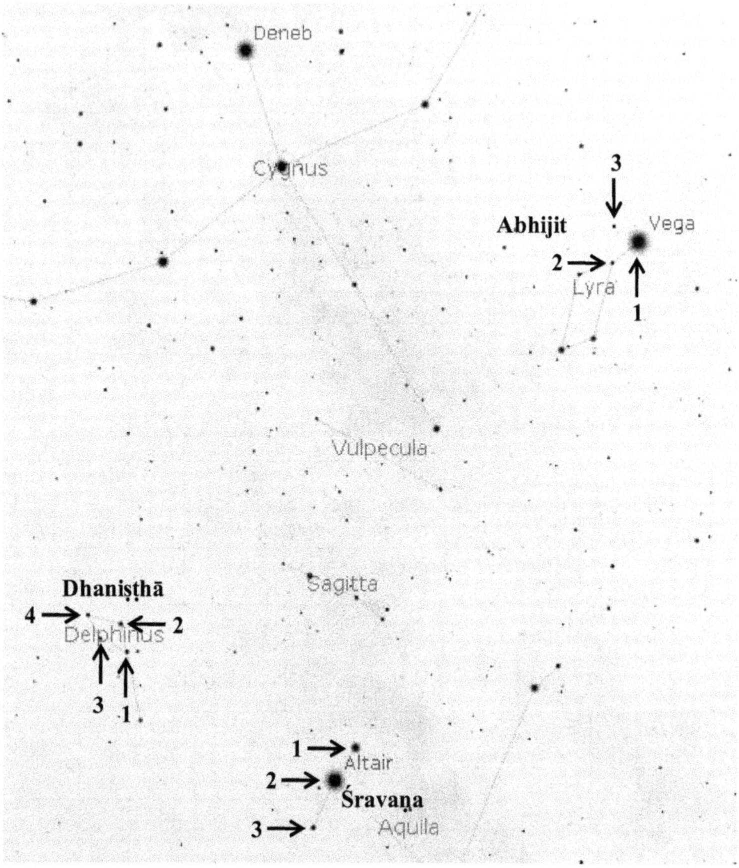

Figure 7.9: Abhijit, Śravaṇa, and Dhaniṣṭhā nakṣatras

Table 7.24: Abhijit nakṣatra

	Proper name	Bayer designation	Flamsteed designation	Apparent Magnitude	Ecliptic longitude	Ecliptic latitude
1.	**Vega**	**α Lyr**	**3 Lyr**	**0.00**	**285° 19′**	**61° 44′**
2.	Nasr Alwaki I	ζ1 Lyr	6 Lyr	4.30	288° 06′	60° 21′
3.	Double Double I	ε1 Lyr	4 Lyr	5.00	288° 38′	62° 24′

Table 7.25: Śravaṇa nakṣatra

	Proper name	Bayer designation	Flamsteed designation	Apparent Magnitude	Ecliptic longitude	Ecliptic latitude
1.	Tarazed	γ Aql	50 Aql	2.70	300° 56′	31° 15′
2.	**Altair**	**α Aql**	**53 Aql**	**0.75**	**301° 47′**	**29° 18′**
3.	Alshain	β Aql	60 Aql	3.70	302° 25′	26° 40′

Table 7.26: Dhaniṣṭhā nakṣatra

	Proper name	Bayer designation	Flamsteed designation	Apparent Magnitude	Ecliptic longitude	Ecliptic latitude
1.	**Rotanev**	**β Del**	**6 Del**	**4.10**	**316° 20′**	**31° 55′**
2.	Sualocin	α Del	9 Del	3.85	317° 23′	33° 01′
3.	Al Ukud	δ Del	11 Del	4.40	318° 07′	31° 57′
4.	Al Salib	γ2 Del	12 Del	4.25	319° 22′	32° 42′

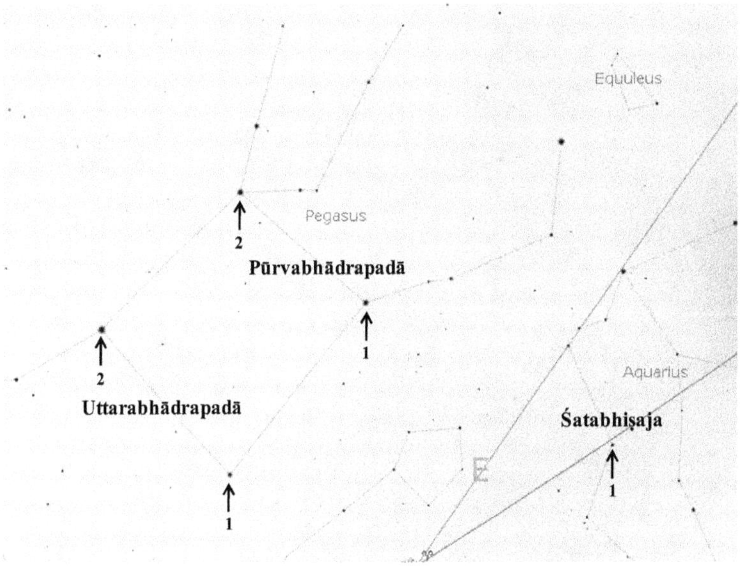

Figure 7.10: Śatabhiṣaja, Pūrvabhādrapadā, and Uttarabhādrapadā nakṣatras

Table 7.27: Śatabhiṣaja nakṣatra

	Proper name	Bayer designation	Flamsteed designation	Apparent Magnitude	Ecliptic longitude	Ecliptic latitude
1.	**Hydor**	**λ Aqr**	**73 Aqr**	**3.70**	**341° 35′**	**-0° 23′**

Table 7.28: Pūrva-bhādrapadā nakṣatra

	Proper name	Bayer designation	Flamsteed designation	Apparent Magnitude	Ecliptic longitude	Ecliptic latitude
1.	**Markab**	**α Peg**	**54 Peg**	**2.45**	**353° 29′**	**19° 24′**
2.	Scheat	β Peg	53 Peg	2.40	359° 22′	31° 08′

Table 7.29: Uttara-bhādrapadā nakṣatra

	Proper name	Bayer designation	Flamsteed designation	Apparent Magnitude	Ecliptic longitude	Ecliptic latitude
1.	Algenib	γ Peg	88 Peg	2.80	9° 09′	12° 36′
2.	**Alpheratz**	**α And** **δ Peg**	**21 And**	**2.05**	**14° 19′**	**25° 41′**

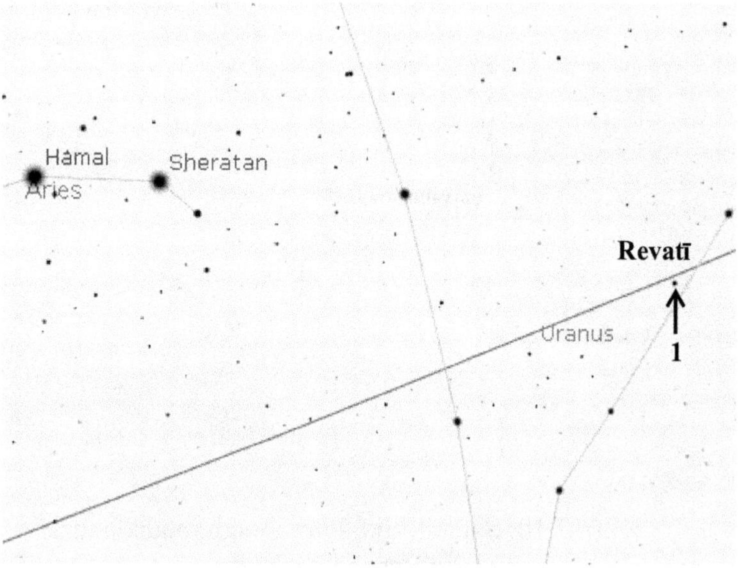

Figure 7.11: Revatī nakṣatra

Table 7.30: Revatī nakṣatra

	Proper name	Bayer designation	Flamsteed designation	Apparent Magnitude	Ecliptic longitude	Ecliptic latitude
1.	**Revati (Kuton II)**	**ζ Psc A**	**86 Psc A**	**5.20**	**19° 53′**	**-0° 13′**

With the tabulation of currently accepted identifications of nakṣatra star groups and yogatārās and their ecliptic coordinates, these identifications can be reassessed to check if other stars fit the description in the astronomical texts better. For this purpose, a precise determination of the boundaries of nakṣatras is required.

Notes

1. Burgess (1860)
2. Pingree and Morrissey (1989)
3. Kaye (1924)

"He insisted that stars were people so well loved, they were traced in constellations, to live forever"
— Jodi Picoult

8. Determination of Nakṣatra Boundaries

Though later astronomical texts divide the celestial sphere in 27 nakṣatras, earlier texts also describe a system consisting of 28 nakṣatra. It takes the moon ~27.32 days to return to the same position among the stars. Based on the observation that the sidereal month is more than 27 days but less than 28 days, the path of moon in the background of the stars was divided in 27 or 28 divisions. Atharvaveda Saṃhitā 19.7.1-5 lists 28 nakṣatra beginning with Kṛttikā and ending in Bharaṇī. Taittirīya Saṃhitā iv.4.10 and Taittirīya Brāhmaṇa 1.5.2.7 list 27 nakṣatra beginning with Kṛttikā and ending in Bharaṇī. Nakṣatra Abhijit is part of the list of 28 nakṣatras, but is dropped from the list in the system of 27 nakṣatras. There is a dialogue between gods Indra and Skanda regarding the dropping of nakṣatra Abhijit in Mahābhārata (Vana Parva, 230.8-10). In this dialogue, Indra says to Skanda that because of jealousy with Rohiṇī, her younger sister Abhijit has gone to forest to do penance. The list of nakṣatra in Sūrya Siddhānta begins with Aświnī. A Jain astronomical text Sūrya Prajnapti

(10.1) gives five other systems besides the one followed by Jains, which started from Abhijit and ended at Uttarāṣāḍhā. These five systems were: 1. Kṛttikā to Bharaṇī; 2. Maghā to Āśleṣā; 3. Dhaniṣṭhā to Śravaṇa; 4. Aświnī to Revatī; and 5. Bharaṇī to Aświnī.

According to Yajuṣ Vedāṅga Jyotiṣa (Verse 7), sun was at the first point of the Dhaniṣṭhā nakṣatra on the day of winter solstice. This suggests that the system of Dhaniṣṭhā as the first nakṣatra was based on sun being at the beginning of Dhaniṣṭhā nakṣatra on winter solstice. Sun was in Maghā nakṣatra on summer solstice around the same time when it was in Kṛttikā nakṣatra on vernal equinox. Thus different systems of nakṣatras were related to the careful observation of solstices and equinoxes. A story in Mahābhārata not only shows the importance of winter solstice but also the desire of the writers of Mahābhārata to carry forward this knowledge to future generations. The story is that of the death of one of the most beloved characters of Mahābhārata, Bhīṣma, and is told in Bhīṣma Parva (120.51-53). According to this story, when Bhīṣma is incapacitated on the battlefield, he refuses to die. He says that he will lie on the bed of arrows till the time of winter solstice. When sun starts its northward journey, only then he will leave this world. He waited for close to two months for winter solstice to take place and then left this world. This story has been passed on from generation to generation, and the dramatic nature of this narrative ensures that the listener will know the definition of winter solstice, which is the day when sun starts its northward journey. To make sure that the message gets passed on to future generations, a very dramatic situation was created in the storyline. From the details of the story, it is clear that the event cannot be historic as no one can control his time of death and lying on a bed of arrows for close to two months is an improbable event. What the story tells us is that winter solstice, and by implication summer solstice and equinoxes, were being carefully observed in India for many millennia. It would have been obvious to Indian astronomers that the position of sun among the stars during solstices and equinoxes was slowly changing due to the effect of precession. When the change in the position of sun became significant, the order of nakṣatras in the list was revised to reflect the new position of the sun on vernal equinox. Out of the different systems mentioned above, the nakṣatra systems with orders Kṛttikā to Bharaṇī, Bharaṇī to Aświnī, and Aświnī to Revatī are important for correct interpretation of coordinates of yogatārās given in Sūrya Siddhānta.

8.1: Rohiṇī system

Based on my analysis of the coordinates of yogatārās given in Sūrya Siddhānta, I have shown in my recently published peer-reviewed paper that the yogatārā of Rohiṇī was at zero longitude of the original nakṣatra system [1]. I have named this system with origin at the yogatārā of Rohiṇī as the Rohiṇī system. The yogatārā of Rohiṇī according to Table 7.6 is Aldebaran (α Tau). The Rohiṇī system with yogatārā of Rohiṇī, Aldebaran, at zero longitude is shown in Figures 8.1a to 8.1d. In this system, the yogatārā of Rohiṇī, Aldebaran, is at the boundary of Rohiṇī and Kṛttikā. The ecliptic longitudes of yogatārās, as identified by Burgess (1860), in Rohiṇī system are shown in Table 8.1 along with the longitudes (dhruvaka) of yogatārās given in Sūrya Siddhānta 8.1-9. The numbers in Figures 8.1a-d refer to the serial number of nakṣatras shown in Table 8.1. The ecliptic longitudes in Rohiṇī system were obtained by setting a date on which the ecliptic longitude of the yogatārā of Rohiṇī, Aldebaran, became 0° 0′. This date was found to be June 12, -3044 by trial and error using Stellarium software. From Table 8.1, it is seen that the longitude of the yogatārā of Rohiṇī had shifted by approximately 50° from its zero point in Rohiṇī system during the time of writing of Sūrya Siddhānta. If the ecliptic longitudes were updated by shifting the zero point of longitude to the beginning of Revatī nakṣatra, which is fourth nakṣatra from Rohiṇī, then 53° 20′ should have been added to the ecliptic longitude in Rohiṇī system instead of 50°. This raises the possibility that another system was also in use, and there was confusion between these two systems resulting in a mix of data derived from two different systems.

8.2: Kṛttikā system

While Sūrya Siddhānta gives 180° as the longitude of Citrā (see Table 5.1), Paitāmaha Siddhānta gives 183° [2]. This could simply be a result of using coordinate systems having boundaries 3° 20′ apart. From Table 8.1, it is seen that the difference between ecliptic longitudes of the yogatārās of Rohiṇī and Kṛttikā is approximately 10° or three quarters of a nakṣatra. A nakṣatra system with its origin at the yogatārā of Kṛttikā will have nakṣatra boundaries at 10° or 3° 20′ from the nakṣatra boundaries in the Rohiṇī system as span of each nakṣatra is 13° 20′. I have named the system with origin at the yogatārā of Kṛttikā as the Kṛttikā system [1].

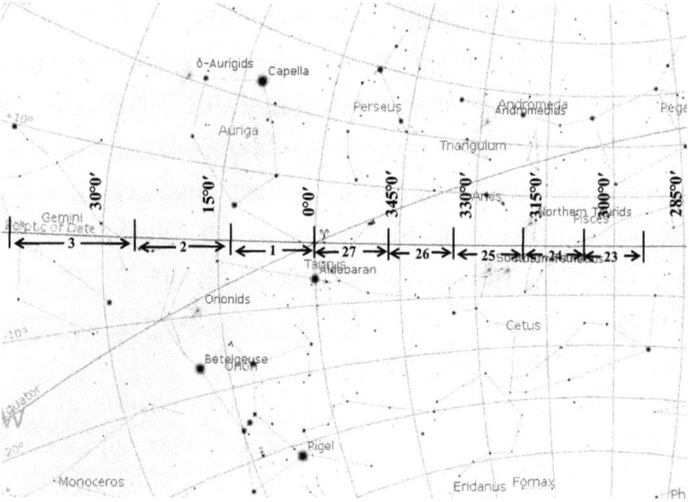

Figure 8.1a: Rohiṇī system

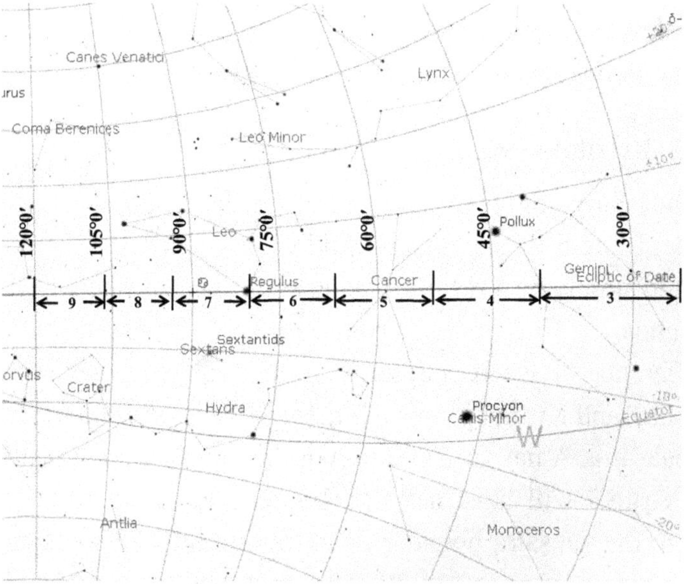

Figure 8.1b: Rohiṇī system (continued)

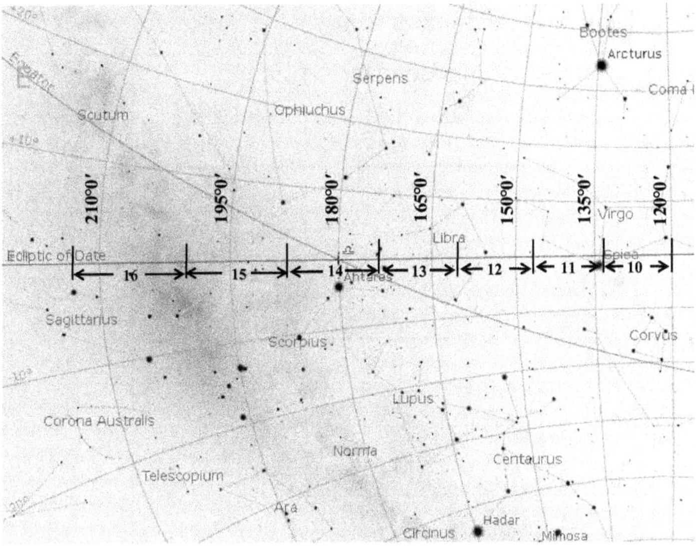

Figure 8.1c: Rohiṇī system (continued)

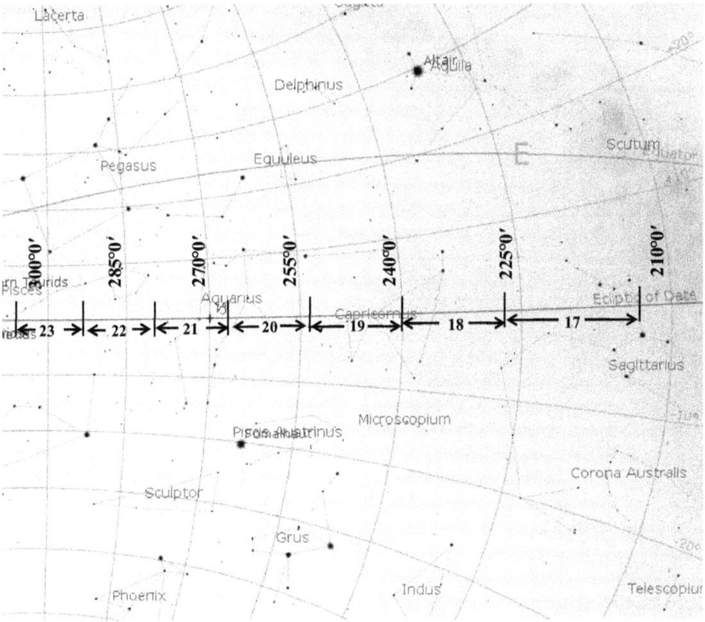

Figure 8.1d: Rohiṇī system (continued)

Table 8.1: Longitudes of yogatārās in Rohiṇī system

No.	Nakṣatra	Junction-star (Yogatārā)*	Ecliptic longitude**	Dhruvaka (longitude)***
1	Rohiṇī	α Tau	0° 00′	49° 30′
2	Mṛgaśirā	λ Ori	13° 58′	63° 0′
3	Ārdrā	α Ori	18° 58′	67° 20′
4	Punarvasu	β Gem	44° 20′	93° 0′
5	Puṣya	δ Cnc	58° 56′	106° 0′
6	Āśleṣā	ε Hya	63° 02′	109° 0′
7	Maghā	α Leo	80° 29′	129° 0′
8	Pūrva-phālgunī	δ Leo	91° 11′	144° 0′
9	Uttara-phālgunī	β Leo	102° 20′	155° 0′
10	Hasta	δ Crv	124° 04′	170° 0′
11	Citrā	α Vir	134° 11′	180° 0′
12	Swāti	α Boo	134° 10′	199° 0′
13	Viśākhā	ι Lib	161° 20′	213° 0′
14	Anurādhā	δ Sco	172° 52′	224° 0′
15	Jyeṣṭhā	α Sco	180° 05′	229° 0′
16	Mūla	λ Sco	194° 54′	241° 0′
17	Pūrvāṣāḍhā	δ Sgr	204° 50′	254° 0′
18	Uttarāṣāḍhā	σ Sgr	212° 40′	260° 0′
	Abhijit	α Lyr	215° 22′	266° 40′
19	Śravaṇa	α Aql	231° 14′	280° 0′
20	Dhaniṣṭhā	β Del	246° 44′	290° 0′
21	Śatabhiṣaja	λ Aqr	271° 49′	320° 0′
22	Pūrva-bhādrapadā	α Peg	283° 57′	326° 0′
23a	Uttara-bhādrapadā	α And	304° 51′	337° 0′
23b	Uttara-bhādrapadā	γ Peg	299° 36′	337° 0′

Table 8.1: Longitudes of yogatārās in Rohiṇī system (continued)

No.	Nakṣatra	Junction-star (Yogatārā)*	Ecliptic longitude**	Dhruvaka (longitude)***
24	Revatī	ζ Psc	310° 01′	359° 50′
25	Aświnī	β Ari	324° 15′	8° 0′
26	Bharaṇī	35 Ari	337° 19′	20° 0′
27	Kṛttikā	η Tau	350° 18′	37° 30′

* As identified by Burgess (1860).
** Ecliptic longitudes on June 12, -3044 based on Stellarium software.
*** Longitudes as given in Sūrya Siddhānta 8.1-9.

The yogatārā of Kṛttikā according to Table 7.5 is Alcyone (η Tau). In Kṛttikā system, the yogatārā of Kṛttikā, Alcyone, is placed at the boundary between Rohiṇī and Kṛttikā. This system is shown in Figures 8.2a-d. In Kṛttikā system, Alcyone is at the end of Kṛttikā nakṣatra. The ecliptic longitudes of yogatārās, as identified by Burgess in 1860, in Kṛttikā system is shown in Table 8.2. The numbers in Figures 8.2a-d refer to the serial number of nakṣatras shown in Table 8.2. The ecliptic longitudes in Kṛttikā system were obtained by setting a date on which the ecliptic longitude of the yogatārā of Kṛttikā, Alcyone, became 0° 0′. This date was found to be April 17, -2336 by trial and error using Stellarium software.

Some observations can be made regarding the nakṣatra boundaries in Rohiṇī system and Kṛttikā system from the information presented in Figures 8.1 and 8.2 and Tables 8.1 and 8.2. According to Sūrya Siddhānta, yogatārā of Revatī (ζ Piscium) is near the origin of the coordinate system. However, ζ Piscium is a very dim star with an apparent magnitude of 5.20.

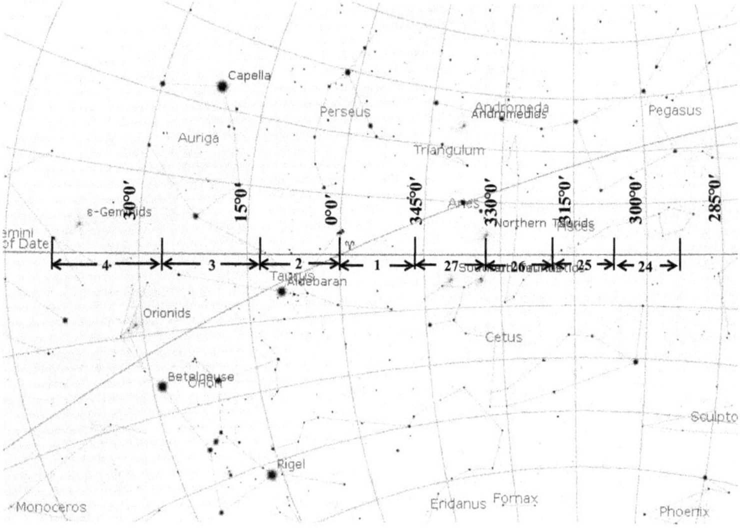

Figure 8.2a: Kṛttikā system

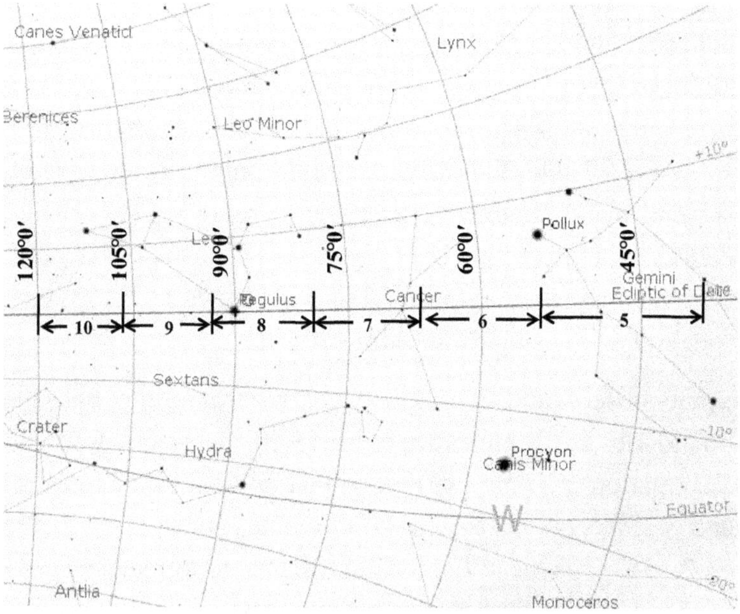

Figure 8.2b: Kṛttikā system (continued)

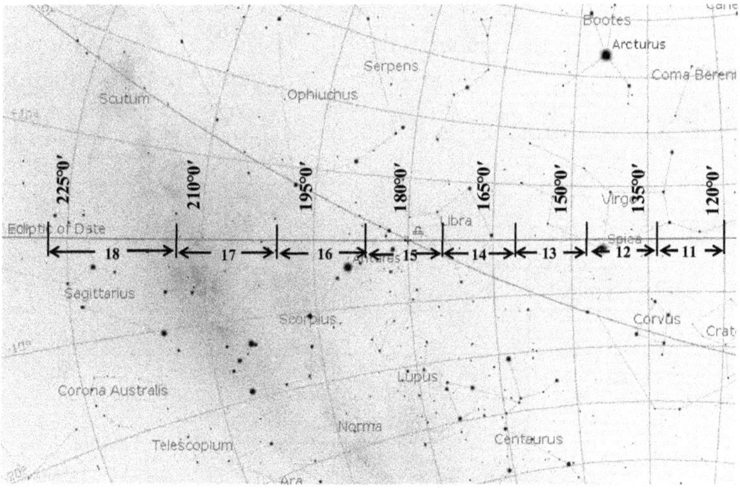

Figure 8.2c: Kṛttikā system (continued)

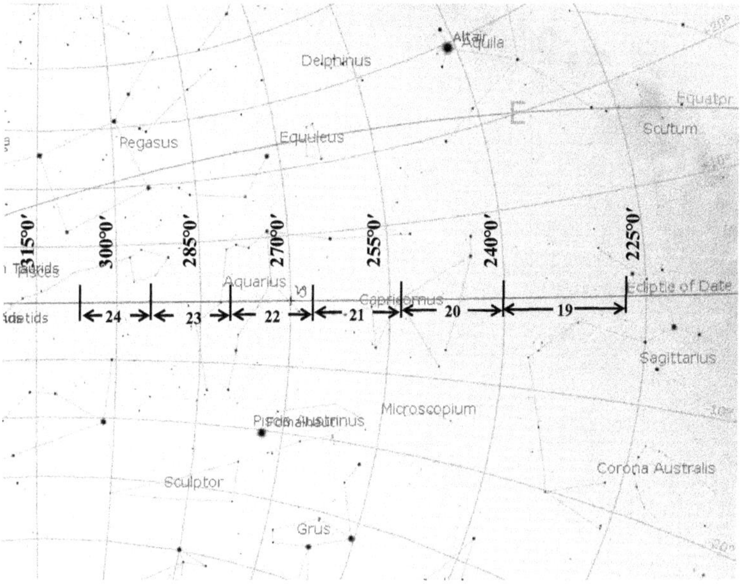

Figure 8.2d: Kṛttikā system (continued)

Table 8.2: Longitudes of yogatārās in Kṛttikā system

No.	Nakṣatra	Junction-star (Yogatārā)*	Ecliptic longitude**	Dhruvaka (longitude)***
1	Kṛttikā	η Tau	0° 00′	37° 30′
2	Rohiṇī	α Tau	9° 43′	49° 30′
3	Mṛgaśirā	λ Ori	23° 40′	63° 0′
4	Ārdrā	α Ori	28° 41′	67° 20′
5	Punarvasu	β Gem	53° 55′	93° 0′
6	Puṣya	δ Cnc	68° 39′	106° 0′
7	Āśleṣā	ε Hya	72° 41′	109° 0′
8	Maghā	α Leo	90° 08′	129° 0′
9	Pūrva-phālgunī	δ Leo	100° 56′	144° 0′
10	Uttara-phālgunī	β Leo	111° 59′	155° 0′
11	Hasta	δ Crv	133° 43′	170° 0′
12	Citrā	α Vir	143° 53′	180° 0′
13	Swāti	α Boo	143° 56′	199° 0′
14	Viśākhā	ι Lib	171° 02′	213° 0′
15	Anurādhā	δ Sco	182° 34′	224° 0′
16	Jyeṣṭhā	α Sco	189° 47′	229° 0′
17	Mūla	λ Sco	204° 36′	241° 0′
18	Pūrvāṣāḍhā	δ Sgr	214° 32′	254° 0′
19	Uttarāṣāḍhā	σ Sgr	222° 22′	260° 0′
	Abhijit	α Lyr	225° 08′	266° 40′
20	Śravaṇa	α Aql	241° 03′	280° 0′
21	Dhaniṣṭhā	β Del	256° 26′	290° 0′
22	Śatabhiṣaja	λ Aqr	281° 32′	320° 0′
23	Pūrva-bhādrapadā	α Peg	293° 37′	326° 0′

Table 8.2: Longitudes of yogatārās in Kṛttikā system (continued)

No.	Nakṣatra	Junction-star (Yogatārā)*	Ecliptic longitude**	Dhruvaka (longitude)***
24a	Uttara-bhādrapadā	α And	314° 31′	337° 0′
24b	Uttara-bhādrapadā	γ Peg	309° 17′	337° 0′
25	Revatī	ζ Psc	319° 44′	359° 50′
26	Aświnī	β Ari	333° 57′	8° 0′
27	Bharaṇī	35 Ari	347° 00′	20° 0′

* As identified by Burgess (1860).
** Ecliptic longitudes on April 17, -2336 based on Stellarium software.
*** Longitudes as given in Sūrya Siddhānta 8.1-9.

Why would such a star be chosen at the origin? Pingree and Morrissey write:

> It is disturbing that ζ Piscium is so dim, and that its α is $0;7^h$ or nearly 2° too high on the assumption that the original list was drawn up in A.D. 425, though the situation, of course, improves as one increases that date. But there are no other visible stars in the neighbourhood. [2]

The yogatārā of Revatī was not at the origin of the coordinate system, when it was designed. The yogatārā of Rohiṇī, Aldebaran, was at the origin of the coordinate system, which is a bright star with apparent magnitude of 0.85. The yogatārā of Revatī was not even a boundary star in the original system. It became a boundary star in the Kṛttikā system devised later. According to Table 8.2, the longitude of the yogatārā of Revatī is 319° 44′ in the Kṛttikā system, which is close to the nakṣatra boundary at

320°. The yogatārā of Punarvasu, Pollux, also became a boundary star in the Kṛttikā system. According to Table 8.2, the longitude of the yogatārā of Punarvasu is 53° 55′ in the Kṛttikā system, which is close to the nakṣatra boundary at 53° 20′.

The yogatārā of Jyeṣṭhā, Antares, is at 180° in the Rohiṇī system according to Table 8.1. When vernal equinox was at the yogatārā of Rohiṇī, then autumnal equinox was at the yogatārā of Jyeṣṭhā.

The yogatārā of Maghā, Regulus, is a boundary star in the Rohiṇī system. Its longitude is 80° 29′ according to Table 8.1, which is close to the nakṣatra boundary at 80° in the Rohiṇī system. Its longitude is 90° 08′ in the Kṛttikā system according to Table 8.2. Thus when vernal equinox fell on the yogatārā of Kṛttikā, the summer solstice fell on the yogatārā of Maghā.

With the precise determination of the nakṣatra boundaries, coordinates given in astronomical texts can be better analyzed to check whether some of the yogatārās have been wrongly identified.

Notes

1. Roy (2019)
2. Pingree and Morrissey (1989).

"It is not until you change your identity to match your life blueprint that you will understand why everything in the past never worked."

—Shannon L. Alder

9. Reassessing the Identification of Nakṣatra Stars

The longitude and latitude data of yogatārās given in Sūrya Siddhānta and other astronomical texts are currently accepted as polar coordinates. For the identification of yogatārās, the assumed polar coordinates were converted to ecliptic coordinates by Burgess [1]. The ecliptic latitude of a star does not change to a significant extent with time unless the proper motion of star is large. Since none of the stars identified as yogatārās have high proper motion, their ecliptic latitudes have not changed significantly over last five millennia and the changes in latitudes over this long period are well within the margin of error expected from ancient and medieval astronomers. Thus ecliptic latitudes can be compared regardless of when the ecliptic latitudes were measured. Table 9.1 shows the latitudes given in Sūrya Siddhānta, corresponding to converted ecliptic latitudes by Burgess [1] and ecliptic latitudes of identified yogatārās in 2000 CE.

Table 9.1: Comparison of latitudes given in Sūrya Siddhānta with ecliptic latitudes calculated by Burgess [1]

No.	Nakṣatra	Yogatārā	Vikṣepa (latitude)*	Converted ecliptic latitude**	Ecliptic latitude***
1	Aświnī	β Ari	10° 0′ N	9° 11′ N	8° 29′ N
2	Bharaṇī	35 Ari	12° 0′ N	11° 06′ N	11° 19′ N
3	Kṛttikā	η Tau	5° 0′ N	4° 44′ N	4° 03′ N
4	Rohiṇī	α Tau	5° 0′ S	4° 49′ S	5° 28′ S
5	Mṛgaśirā	λ Ori	10° 0′ S	9° 49′ S	13° 22′ S
6	Ārdrā	α Ori	9° 0′ S	8° 53′ S	16° 02′ S
7	Punarvasu	β Gem	6° 0′ N	6° 0′ N	6° 41′ N
8	Puṣya	δ Cnc	0° 0′	0° 0′ N	0° 05′ N
9	Āśleṣā	ε Hya	7° 0′ S	6° 56′ S	11° 06′ S
10	Maghā	α Leo	0° 0′	0° 0′	0° 28′ N
11	Pūrva-phālgunī	δ Leo	12° 0′ N	11° 19′ N	14° 20′ N
12	Uttara-phālgunī	β Leo	13° 0′ N	12° 5′ N	12° 16′ N
13	Hasta	δ Crv	11° 0′ S	10° 6′ S	12° 12′ S
14	Citrā	α Vir	2° 0′ S	1° 50′ S	2° 03′ S
15	Swāti	α Boo	37° 0′ N	33° 50′ N	30° 44′ N
16	Viśākhā	ι Lib	1°30′ S	1°25′ N	1°51′ S
17	Anurādhā	δ Sco	3° 0′ S	2° 52′ S	1° 59′ S
18	Jyeṣṭhā	α Sco	4° 0′ S	3° 50′ S	4° 34′ S
19	Mūla	λ Sco	9° 0′ S	8° 48′ S	13° 47′ S
20	Pūrvāṣāḍhā	δ Sgr	5° 30′ S	5° 28′ S	6° 28′ S
21	Uttarāṣāḍhā	σ Sgr	5° 0′ S	4° 59′ S	3° 27′ S
22	Abhijit	α Lyr	60° 0′ N	59° 58′ N	61° 44′ N
23	Śravaṇa	α Aql	30° 0′ N	29° 54′ N	29° 18′ N
24	Dhaniṣṭhā	β Del	36° 0′ N	35° 33′ N	31° 55′ N
25	Śatabhiṣaja	λ Aqr	0° 30′ S	0° 28′ S	0° 23′ S

Table 9.1 (continued)

No.	Nakṣatra	Yogatārā	Vikṣepa (latitude)*	Converted ecliptic latitude**	Ecliptic latitude***
26	Pūrva-bhādrapadā	α Peg	24° 0′ N	22° 30′ N	19° 24′ N
27	Uttara-bhādrapadā	α And	26° 0′ N	24° 1′ N	25° 41′ N
28	Revatī	ζ Psc	0° 0′	0° 0′ S	0° 13′ S

* Latitudes as given in Sūrya Siddhānta 8.1-9.
** As calculated by Burgess [1] assuming that Sūrya Siddhānta provides polar latitude.
*** J2000.0 ecliptic latitudes based on Stellarium software.

In 25 cases out of 28, the latitudes given in Sūrya Siddhānta are within 1° of the converted ecliptic latitudes by Burgess [1]. In 16 cases out of 28, the given latitudes in Sūrya Siddhānta provide an equal or better fit to the actual ecliptic latitudes compared to the converted ecliptic latitudes. This shows that as far as latitude is concerned, there is no basis for the assumption that the latitudes given in Sūrya Siddhānta are polar latitudes.

The question then is if the coordinates given in Sūrya Siddhānta are ecliptic coordinates, would not it be obvious? If the longitudes given are ecliptic longitudes, then all the given longitudes will be off by same amount from the current ecliptic longitudes due to change in zero point from which ecliptic longitudes are measured. The difference between current ecliptic longitudes and those given in Sūrya Siddhānta varies from 5° for Swāti to 37° for Uttara-bhādrapadā. This has resulted in researchers looking at different time periods when the measurements were made.

Since no single time period can be found where the data gives a good fit and Indian astronomers keep repeating the same data over several centuries, doubts have been raised about the ability of Indian astronomers to measure coordinates.

It is clear that Indian astronomers did not keep measuring the coordinates because they believed the coordinates to be ecliptic coordinates which do not change when fixed to stars in a sidereal system. The coordinates of the yogatārā of Aświnī are given as 8° longitude and 10° latitude in Sūrya Siddhānta, Paitāmaha Siddhānta, Mahābhāskarīya and Laghubhāskarīya, and by Brahmagupta, Vaṭeśvara, Lalla, and Gaṇeśa [2]. It was never changed as would be expected in a sidereal ecliptic coordinate system. The reason that the data given in texts do not fit to a single time period is due to a mix up of data during the last update when the zero point was changed to account for precession as well as misidentification of some yogatārās. The data given for each yogatārā is examined next to clarify which measurement system the data belongs to and whether the yogatārā has been correctly identified.

9.1: The yogatārā of Aświnī

Since Aświnī is the first nakṣatra in the list of nakṣatras given in Sūrya Siddhānta and other texts of classical period, it is of vital importance to correctly identify the yogatārā of Aświnī. As mentioned above, every text gives the coordinates of the yogatārā of Aświnī as 8° longitude and 10° latitude. As shown in Figure 7.1 and Table 7.3, Aświnī has two stars in its star group, which are currently identified as Sheratan (β Ari) and Mesarthim (γ1 Ari). Out of these two stars, currently accepted yogatārā of Aświnī is Sheratan (β Ari) with apparent magnitude of 2.60 and J2000.0 ecliptic longitude and latitude of 33° 58′ and 8° 29′ respectively. Figure 7.1 also shows the star Hamal (α Ari), which is the brightest star of the Aries constellation. Hamal (α Ari) has apparent magnitude of 2.00 and J2000.0 ecliptic longitude and latitude of 37° 40′ and 9° 58′ respectively. The ecliptic latitude of 9° 58′ of Hamal matches closely with the latitude of 10° given in Sūrya Siddhānta. This raises the possibility that the yogatārā of Aświnī is Hamal instead of Sheratan. This is confirmed from the longitude of 8° given in every text. The ecliptic longitude of Hamal is 327° 50′ in the Rohiṇī system described in section 8.1. That is when the ecliptic longitude of Rohiṇī was 0° 0′ on June 12, -3044, the ecliptic longitude of Hamal was 327° 50′ according to Stellarium software. Since Aświnī is three nakṣatras away from Rohiṇī, the span of Aświnī nakṣatra in Rohiṇī system is 320° 0′ to 333° 20′. This means that Hamal is 7° 50′ away from the beginning of Aświnī nakṣatra in Rohiṇī system and when nakṣatra list was updated to begin with Aświnī, Hamal would have ecliptic longitude of 7° 50′ in the new system. This is a close match with 8° longitude of the yogatārā of Aświnī given in all astronomical texts. Thus Hamal has excellent match for both latitude and longitude to be the yogatārā of Aświnī. Figure 9.1 shows the proposed and accepted yogatārās of Aświnī nakṣatra. The identification of Hamal as the yogatārā of Aświnī with 8° longitude from the beginning of Aświnī provides a zero point for Vedic-Hindu astronomy.

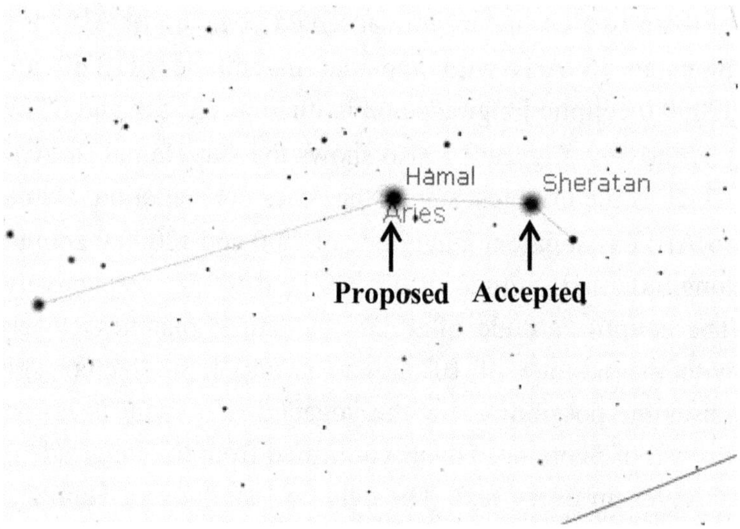

Figure 9.1: Proposed and accepted yogatārās of Aświnī nakṣatra

The date on which the ecliptic longitude of Hamal was 8° 0′ is found to be April 13, -130 by trial and error using Stellarium software. Table 9.2 shows the ecliptic longitude of yogatārās as identified by Burgess (1860), when Hamal had an ecliptic longitude of 8° 0′. I have named this system Aświnī-beginning Rohiṇī system as it measures the longitude from the beginning of Aświnī nakṣatra in the Rohiṇī system. Another system that is important for reassessing the identification of yogatārās is Aświnī-beginning Kṛttikā system, which measures the longitude from the beginning of Aświnī nakṣatra in Kṛttikā system. Rohiṇī and Kṛttikā systems have been defined in the previous chapter.

Table 9.2: Longitude of yogatārās in Aświnī-beginning Rohiṇī system

No.	Nakṣatra	Yogatārā*	Relative Longitude
1	Aświnī	β Ari	4° 21′
2	Bharaṇī	35 Ari	17° 22′
3	Kṛttikā	η Tau	30° 23′
4	Rohiṇī	α Tau	40° 09′
5	Mṛgaśirā	λ Ori	54° 05′
6	Ārdrā	α Ori	59° 07′
7	Punarvasu	β Gem	83° 57′
8	Puṣya	δ Cnc	99° 05′
9	Āśleṣā	ε Hya	102° 55′
10	Maghā	α Leo	120° 22′
11	Pūrva-phālgunī	δ Leo	131° 31′
12	Uttara-phālgunī	β Leo	142° 11′
13	Hasta	δ Crv	163° 59′
14	Citrā	α Vir	174° 15′
15	Swāti	α Boo	174° 30′
16	Viśākhā	ι Lib	201° 25′
17	Anurādhā	δ Sco	212° 58′
18	Jyeṣṭhā	α Sco	220° 10′
19	Mūla	λ Sco	234° 59′
20	Pūrvāṣāḍhā	δ Sgr	244° 57′
21	Uttarāṣāḍhā	σ Sgr	252° 46′
22	Abhijit	α Lyr	255° 40′
23	Śravaṇa	α Aql	271° 50′
24	Dhaniṣṭhā	β Del	286° 47′
25	Śatabhiṣaja	λ Aqr	311° 57′
26	Pūrva-bhādrapadā	α Peg	323° 57′
27a	Uttara-bhādrapadā	α And	344° 48′
27b	Uttara-bhādrapadā	γ Peg	339° 36′
28	Revatī	ζ Psc	350° 12′

*as identified by Burgess [1]

Table 9.3 shows the ecliptic longitude of yogatārās as identified by Burgess [1] in the Aświnī-beginning Kṛttikā system. The longitudes have been calculated by setting the longitude of the yogatārā of Revatī (ζ Piscium) to 359° 50′, which is the longitude given in Sūrya Siddhānta. According to Table 8.2, the ecliptic longitude of the yogatārā of Revatī (ζ Piscium) is 319° 44′ in Kṛttikā system, which means the yogatārā of Revatī is a boundary star in Kṛttikā system as the nakṣatra boundaries are at 13° 20′ intervals, and hence one of the boundaries is at 320°. Since the boundary of Aświnī and Revatī nakṣatras is at 40° from the boundary of Rohiṇī and Kṛttikā nakṣatras, the coordinate of the yogatārā of Revatī is 359° 44′ in the Aświnī-beginning Kṛttikā system, which is an excellent match with the value of 359° 50′ given in Sūrya Siddhānta. The date on which the ecliptic longitude of Revatī star (ζ Piscium) was 359° 50′ is found to be December 12, 563 by trial and error using Stellarium software.

The two systems of coordinates created a confusion, which resulted in a list that has some coordinates from Aświnī-beginning Rohiṇī system, some coordinates from Aświnī-beginning Kṛttikā system, and some coordinates resulting from confusion between Rohiṇī and Kṛttikā systems. To illustrate this point, a column with title "3° 20′ offset" has been added in Table 9.3. The values in this column are 3° 20′ (one quarter of the span of a nakṣatra) lesser than the values in Aświnī-beginning Kṛttikā system. The nakṣatra boundaries in Rohiṇī and Kṛttikā systems are separated by 10° or 3° 20′. With this understanding, the identification of other yogatārās is assessed below.

Table 9.3: Longitude of yogatārās in Aświnī-beginning Kṛttikā system

No.	Nakṣatra	Yogatārā*	Longitude	3° 20′ Offset
1	Aświnī	β Ari	13° 58′	10° 38′
2	Bharaṇī	35 Ari	26° 58′	23° 38′
3	Kṛttikā	η Tau	40° 00′	36° 40′
4	Rohiṇī	α Tau	49° 46′	46° 26′
5	Mṛgaśirā	λ Ori	63° 42′	60° 22′
6	Ārdrā	α Ori	68° 45′	65° 25′
7	Punarvasu	β Gem	93° 27′	90° 07′
8	Puṣya	δ Cnc	108° 42′	105° 22′
9	Āśleṣā	ε Hya	112° 28′	109° 08′
10	Maghā	α Leo	129° 56′	126° 36′
11	Pūrva-phālgunī	δ Leo	141° 12′	137° 52′
12	Uttara-phālgunī	β Leo	151° 45′	148° 25′
13	Hasta	δ Crv	173° 33′	170° 13′
14	Citrā	α Vir	183° 52′	180° 32′
15	Swāti	α Boo	184° 09′	180° 49′
16	Viśākhā	ι Lib	211° 01′	207° 41′
17	Anurādhā	δ Sco	222° 35′	219° 15′
18	Jyeṣṭhā	α Sco	229° 46′	226° 26′
19	Mūla	λ Sco	244° 35′	241° 15′
20	Pūrvāṣāḍhā	δ Sgr	254° 34′	251° 14′
21	Uttarāṣāḍhā	σ Sgr	262° 23′	259° 03′
22	Abhijit	α Lyr	265° 18′	261° 58′
23	Śravaṇa	α Aql	281° 33′	278° 13′
24	Dhaniṣṭhā	β Del	296° 23′	293° 03′
25	Śatabhiṣaja	λ Aqr	321° 34′	318° 14′
26	Pūrva-bhādrapadā	α Peg	333° 32′	330° 12′
27	Uttara-bhādrapadā	γ Peg	349° 12′	345° 52′
28	Revatī	ζ Psc	359° 50′	356° 30′

*as identified by Burgess [1]

9.2: The yogatārā of Bharaṇī

As shown in Figure 7.1 and Table 7.4, Bharaṇī has three stars in its star group, which are currently identified as Barani II (35 Ari), Bharani (41 Ari) and Barani III (39 Ari). Out of these three stars, currently accepted yogatārā of Bharaṇī is Barani II (35 Ari) with apparent magnitude of 4.65 and J2000.0 ecliptic longitude and latitude of 46° 56′ and 11° 19′ respectively. However, Bharani (41 Ari) is the brightest star among the three and should be the yogatārā of Bharaṇī. As seen in Table 9.2, the longitude of Barani II (35 Ari) in Aświnī-beginning Rohiṇī system is 17° 22′, while the longitude given in Sūrya Siddhānta is 20°. The longitude of Bharani (41 Ari) in Aświnī-beginning Rohiṇī system is 18° 36′, which is a better match with the longitude given in Sūrya Siddhānta. Bharani (41 Ari) has latitude of 10° 27′, which is a reasonable match with 12° latitude given in Sūrya Siddhānta. Bharani (41 Ari) is better suited to be the yogatārā of Bharaṇī due to it being brighter and matching longitude better compared to Barani II. Figure 9.2 shows the proposed and accepted yogatārās of Bharaṇī nakṣatra.

9.3: The yogatārā of Kṛttikā

As shown in Figure 7.2 and Table 7.5, Kṛttikā has six stars in its star group with Alcyone being the accepted yogatārā. Alcyone is the brightest star in the group with apparent magnitude of 2.85 and J2000.0 ecliptic longitude and latitude of 60° 00′ and 4° 03′ respectively. This is a well known group of stars, and there can be no doubt about the star cluster being identified as Kṛttikā and its brightest star Alcyone being the yogatārā of Kṛttikā nakṣatra. As seen in

Table 5.1, the longitude given in Sūrya Siddhānta is 37° 30′, which is 2° 30′ lesser than the relative longitude of 40° 0′ in Aświnī-beginning Kṛttikā system.

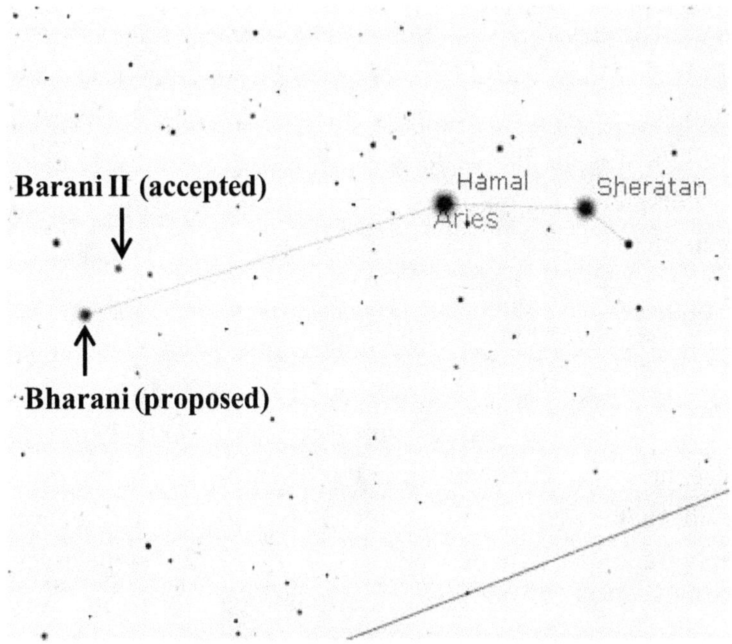

Figure 9.2: The proposed and accepted yogatārās of Bharaṇī nakṣatra

9.4: The yogatārā of Rohiṇī

As shown in Figure 7.3 and Table 7.6, Rohiṇī has five stars in its star group with Aldebaran being the accepted yogatārā. Aldebaran is the brightest star in the group with apparent magnitude of 0.85 and J2000.0 ecliptic longitude and latitude of 69° 47′ and -5° 28′ respectively. This is a well known group of stars, and there can be no doubt about the stars belonging to Rohiṇī nakṣatra and its brightest star

Aldebaran being the yogatārā of Rohiṇī nakṣatra. As seen in Table 5.1, the longitude given in Sūrya Siddhānta is 49° 30′, which is an excellent match with the relative longitude of 49° 46′ in Aświnī-beginning Kṛttikā system. The five stars in Rohiṇī nakṣatra make a configuration of a cart, which is described in Indian texts as the Śakaṭa (cart) of Rohiṇī. Moon had 27 wives according to Purāṇas, whose names are exactly same as the names of 27 nakṣatras. Rohiṇī was his favourite wife, which refers to the frequent occultation of Rohiṇī by moon due to it being close to ecliptic.

9.5: The yogatārā of Mṛgaśirā

As shown in Figure 7.3 and Table 7.7, Mṛgaśirā has three stars in its star group with Meissa (λ Ori) being the accepted yogatārā. Meissa is the brightest star in the group with apparent magnitude of 3.50 and J2000.0 ecliptic longitude and latitude of 83° 42′ and -13° 22′ respectively. As seen in Table 5.1, the longitude given in Sūrya Siddhānta is 63° 0′, which is an excellent match with the relative longitude of 63° 42′ in Aświnī-beginning Kṛttikā system.

9.6: The yogatārā of Ārdrā

As shown in Figure 7.3 and Table 7.8, Ārdrā has only one stars in its star group, Betelgeuse (α Ori), which by default is its yogatārā. Betelgeuse is a very bright star with apparent magnitude of 0.45 and J2000.0 ecliptic longitude and latitude of 88° 45′ and -16° 02′ respectively. As seen in Table 5.1, the longitude given in Sūrya Siddhānta is 67°

20′, which is a reasonable match with the relative longitude of 68° 45′ in Aświnī-beginning Kṛttikā system.

9.7: The yogatārā of Punarvasu

As shown in Figure 7.4 and Table 7.9, Punarvasu has two stars in its star group, which are currently identified as Castor (α Gem) and Pollux (β Gem). Out of these two stars, currently accepted yogatārā of Punarvasu is Pollux (β Gem) with apparent magnitude of 1.15 and J2000.0 ecliptic longitude and latitude of 113° 13′ and 6° 41′ respectively. As seen in Table 5.1, the longitude given in Sūrya Siddhānta is 93° 0′, which is an excellent match with the relative longitude of 93° 27′ in Aświnī-beginning Kṛttikā system.

9.8: The yogatārā of Puṣya

As shown in Figure 7.4 and Table 7.10, Puṣya has three stars in its star group, which are currently identified as HIP 41822 (θ Cnc), Asellus Borealis (γ Cnc), and Asellus Australis (δ Cnc). Out of these three stars, currently accepted yogatārā of Puṣya is Asellus Australis (δ Cnc) with apparent magnitude of 3.90 and J2000.0 ecliptic longitude and latitude of 128° 43′ and 0° 05′ respectively. As seen in Table 5.1, the longitude given in Sūrya Siddhānta is 106° 0′, which is 2° 52′ lesser than the relative longitude of 108° 42′ in Aświnī-beginning Kṛttikā system. However, the value is closer to 3° 20′ offset value of 105° 22′, which means a wrong correction was applied to the longitude of the yogatārā of Puṣya.

9.9: The yogatārā of Āśleṣā

As shown in Figure 7.4 and Table 7.11, Āśleṣā has five stars in its star group with Minazal III (ε Hya) being the accepted yogatārā. Minazal III is the brightest star in the group with apparent magnitude of 3.40 and J2000.0 ecliptic longitude and latitude of 132° 21′ and -11° 06′ respectively. As seen in Table 5.1, the longitude given in Sūrya Siddhānta is 109° 0′, which is 3° 38′ lesser than the relative longitude of 112° 28′ in Aświnī-beginning Kṛttikā system. However, the value is very close to 3° 20′ offset value of 109° 8′, which means a wrong correction was applied to the longitude of Āśleṣā.

9.10: The yogatārā of Maghā

As shown in Figure 7.5 and Table 7.12, Maghā has six stars in its star group with Regulus (α Leo) being the accepted yogatārā. Regulus is the brightest star in the group with apparent magnitude of 1.35 and J2000.0 ecliptic longitude and latitude of 149° 50′ and 0° 28′ respectively. As seen in Table 5.1, the longitude given in Sūrya Siddhānta is 129° 0′, which is a good match with the relative longitude of 129° 56′ in Aświnī-beginning Kṛttikā system.

9.11: The yogatārā of Pūrva-phālgunī

As shown in Figure 7.5 and Table 7.13, Pūrva-phālgunī has two stars in its star group, which are currently identified as Zosma (δ Leo) and Chertan (θ Leo). Out of these two stars, currently accepted yogatārā of Pūrva-phālgunī is Zosma (δ Leo) with apparent magnitude of 2.55 and J2000.0 ecliptic longitude and latitude of 161° 19′ and 14° 20′ respectively.

As seen in Table 5.1, the longitude given in Sūrya Siddhānta is 144° 0′, which is 2° 48′ more than the relative longitude of 141° 12′ in Aświnī-beginning Kṛttikā system. The latitude given in Sūrya Siddhānta is 12° 0′, which matches reasonably with the J2000.0 ecliptic latitude of 14° 20′. There are no other stars in the vicinity that will fit the longitude and latitude better.

9.12: The yogatārā of Uttara-phālgunī

As shown in Figure 7.5 and Table 7.14, Uttara -phālgunī has two stars in its star group, which are currently identified as HIP 57565 (93 Leo) and Denebola (β Leo). Out of these two stars, currently accepted yogatārā of Uttara-phālgunī is Denebola (β Leo) with apparent magnitude of 2.10 and J2000.0 ecliptic longitude and latitude of 171° 37′ and 12° 16′ respectively. As seen in Table 5.1, the longitude given in Sūrya Siddhānta is 155° 0′, which is 3° 15′ more than the relative longitude of 151° 45′ in Aświnī-beginning Kṛttikā system. The latitude given in Sūrya Siddhānta is 13° 0′, which matches very well with the J2000.0 ecliptic latitude of 12° 16′. There are no other stars in the vicinity that will fit the longitude and latitude better.

9.13: The yogatārā of Hasta

As shown in Figure 7.5 and Table 7.15, Hasta has five stars in its star group with Algorab (δ Crv) being the accepted yogatārā with apparent magnitude of 2.90 and J2000.0 ecliptic longitude and latitude of 193° 27′ and -12° 12′ respectively. As seen in Table 5.1, the longitude given in Sūrya Siddhānta is 170° 0′, which is 3° 33′ lesser than the relative longitude of 173° 33′ in Aświnī-beginning Kṛttikā system. It should be noted that Algorab is not the brightest star of the group and in most cases the brightest star of the group has been identified as the yogatārā. The brightest star of the group is Gienah (γ Crv) with apparent magnitude of 2.55 and J2000.0 ecliptic longitude and latitude of 190° 44′ and -14° 30′ respectively. The relative longitude of Gienah is 170° 50′ in Aświnī-beginning Kṛttikā system, which is a good match with the relative longitude of 170° 0′ given in Sūrya Siddhānta. Though the latitude of Algorab is a better match for -11° 0′ latitude given in Sūrya Siddhānta compared to Gienah, on account of brightness and better match of longitude, Gienah has a better claim to be the yogatārā of Hasta nakṣatra. Figure 9.3 shows the proposed and accepted yogatārās of Hasta nakṣatra.

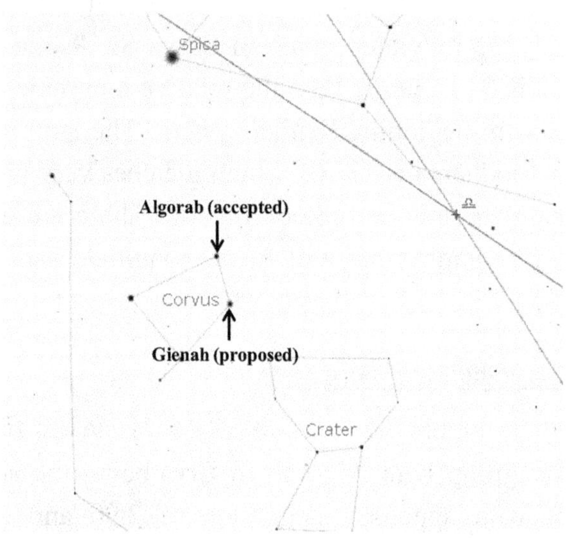

Figure 9.3: The proposed and accepted yogatārās of Hasta nakṣatra

9.14: The yogatārā of Citrā

As shown in Figure 7.6 and Table 7.16, Citrā has only one star in its star group, Spica (α Vir), which by default is its yogatārā. Spica is a very bright star with apparent magnitude of 0.95 and J2000.0 ecliptic longitude and latitude of 203° 50′ and -2° 03′ respectively. As seen in Table 5.1, the longitude given in Sūrya Siddhānta is 180° 0′, which is 3° 52′ lesser than the relative longitude of 183° 52′ in Aświnī-beginning Kṛttikā system. However, the value is close to 3° 20′ offset value of 180° 32′, which means a wrong correction was applied to the longitude of the yogatārā of Citrā.

9.15: The yogatārā of Swāti

As shown in Figure 7.6 and Table 7.17, Swāti has only one star in its star group, Arcturus (α Boo), which by default is its yogatārā. Arcturus is a very bright star with apparent magnitude of 0.15 and J2000.0 ecliptic longitude and latitude of 204° 14′ and 30° 43′ respectively. As seen in Table 5.1, the longitude given in Sūrya Siddhānta is 199° 0′, which is 14° 51′ more than the relative longitude of 184° 9′ in Aświnī-beginning Kṛttikā system. The latitude given in Sūrya Siddhānta is 37° 0′, which is 6° 17′ greater than the J2000.0 ecliptic latitude of 30° 43′ of Arcturus.

The ecliptic longitude of Arcturus is within one degree of the ecliptic longitude of Spica, the yogatārā of previous nakṣatra Citrā. According to Sūrya Siddhānta, the difference in longitudes of the yogatārās of Citrā and Swāti is 19° 0′. Thus Arcturus is not a good fit to be the yogatārā of Swāti. There is another star, Alphecca, which fits the longitude of the yogatārā of Swāti better. Alphecca (α CrB) has apparent magnitude of 2.20 and J2000.0 ecliptic longitude and latitude of 222° 18′ and 44° 19′ respectively. In terms of latitude, Alphecca is approximately 7° higher and Arcturus is approximately 7° lower. The relative longitude of Alphecca is 202° 5′ in Aświnī-beginning Kṛttikā system, which is 3° 5′ greater than the longitude of 199° 0′ given in Sūrya Siddhānta. The longitude matches well with Aświnī-beginning Kṛttikā offset value of 198° 45′. Thus, Alphecca is a much better match in terms of longitude compared to Arcturus. Figure 9.4 shows the proposed and accepted yogatārās of Swāti nakṣatra.

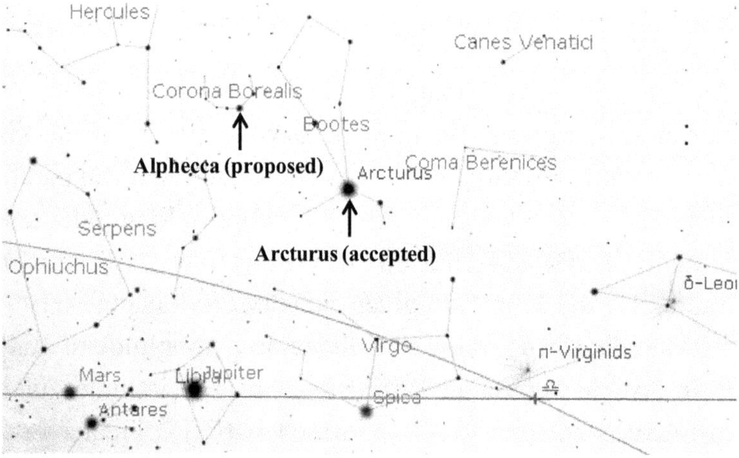

Figure 9.4: The proposed and accepted yogatārās of Swāti nakṣatra.

9.16: The yogatārā of Viśākhā

As shown in Figure 7.6 and Table 7.18, Viśākhā has four stars in its star group with HIP 74392 (ι Lib) being the accepted yogatārā. HIP 74392 is the dimmest star in the group with apparent magnitude of 4.50 and J2000.0 ecliptic longitude and latitude of 231° 0′ and -1° 51′ respectively. As seen in Table 5.1, the longitude given in Sūrya Siddhānta is 213° 0′, which is a reasonable match with the relative longitude of 211° 1′ in Aświnī-beginning Kṛttikā system. The latitude given in Sūrya Siddhānta is -1° 30′, which matches very well with the J2000.0 ecliptic latitude of -1° 51′. It is not clear why a very dim star was chosen as yogatārā, when there are other brighter stars in the group of stars belonging to Viśākhā. However, there is no other star in the vicinity that will fit the longitude and latitude better.

9.17: The yogatārā of Anurādhā

As shown in Figure 7.7 and Table 7.19, Anurādhā has three stars in its star group with Dshubba (δ Sco) being the accepted yogatārā. Dshubba is the brightest star in the group with apparent magnitude of 2.35 and J2000.0 ecliptic longitude and latitude of 242° 34′ and -1° 59′ respectively. As seen in Table 5.1, the longitude given in Sūrya Siddhānta is 224° 0′, which is a good match with the relative longitude of 222° 35′ in Aświnī-beginning Kṛttikā system.

9.18: The yogatārā of Jyeṣṭhā

As shown in Figure 7.7 and Table 7.20, Jyeṣṭhā has three stars in its star group with Antares (α Sco) being the accepted yogatārā. Antares is the brightest star in the group with apparent magnitude of 1.05 and J2000.0 ecliptic longitude and latitude of 249° 46′ and -4° 34′ respectively. As seen in Table 5.1, the longitude given in Sūrya Siddhānta is 229° 0′, which is an excellent match with the relative longitude of 229° 46′ in Aświnī-beginning Kṛttikā system.

9.19: The yogatārā of Mūla

As shown in Figure 7.8 and Table 7.21, Shaula (λ Sco) is the accepted yogatārā of Mūla nakṣatra. Shaula is the brightest star in the group with apparent magnitude of 1.60 and J2000.0 ecliptic longitude and latitude of 264° 35′ and -13° 47′ respectively. As seen in Table 5.1, the longitude given in Sūrya Siddhānta is 241° 0′, which is 3° 35′ lesser than the relative longitude of 244° 35′ in Aświnī-beginning Kṛttikā system. However, the value is very close to 3° 20′ offset value of 241° 15′, which means a wrong correction was applied to the longitude of the yogatārā of Mūla nakṣatra. The latitude given in Sūrya Siddhānta is -9° 0′, which is 4° 47′ more than J2000.0 ecliptic latitude of -13° 47′. There are no other stars in the vicinity that will fit both longitude and latitude better. There can be hardly any doubt that the group of stars listed in Table 7.21 belong to Mūla nakṣatra due to the shape of scorpion's tail made by these stars, which is the shape of Mūla nakṣatra according to Jain astronomical text Jambudvīpaprajñapti 7.192. Only ε Sco out of these stars fits the latitude better, but the match with longitude is not good. Hence, Shaula is the best fit for being the yogatārā of Mūla nakṣatra.

9.20: The yogatārā of Pūrvāṣāḍhā

As shown in Figure 7.8 and Table 7.22, Pūrvāṣāḍhā has two stars in its star group currently identified as Kaus Media (δ Sgr) and Kaus Australis (ε Sgr). Out of these two stars, currently accepted yogatārā of Pūrvāṣāḍhā is Kaus

Media (δ Sgr) with apparent magnitude of 2.70 and J2000.0 ecliptic longitude and latitude of 274° 35′ and -6° 28′ respectively. As seen in Table 5.1, the longitude given in Sūrya Siddhānta is 254° 0′, which is an excellent match with the relative longitude of 254° 34′ in Aświnī-beginning Kṛttikā system. Latitude given in Sūrya Siddhānta is -5° 30′, which matches well with the J2000.0 ecliptic latitude of -6° 28′ of Kaus Media (δ Sgr).

9.21: The yogatārā of Uttarāṣāḍhā

As shown in Figure 7.8 and Table 7.23, Uttarāṣāḍhā has two stars in its star group currently identified as Nunki (σ Sgr) and Ascella (ζ Sgr). Out of these two stars, currently accepted yogatārā of Uttarāṣāḍhā is Nunki (σ Sgr) with apparent magnitude of 2.05 and J2000.0 ecliptic longitude and latitude of 282° 23′ and -3° 27′ respectively. As seen in Table 5.1, the longitude given in Sūrya Siddhānta is 260° 0′, which is 2° 23′ lesser than the relative longitude of 262° 23′ in Aświnī-beginning Kṛttikā system. However, the value is close to 3° 20′ offset value of 259° 03′, which means a wrong correction was applied to the longitude of the yogatārā of Uttarāṣāḍhā.

9.22: The yogatārā of Abhijit

As shown in Figure 7.9 and Table 7.24, Abhijit has three stars in its star group with Vega (α Lyr) being the accepted yogatārā. Vega is the brightest star in the group with apparent magnitude of 0.00 and J2000.0 ecliptic longitude and latitude of 285° 19′ and 61° 44′ respectively. As seen in Table 5.1, the longitude given in Sūrya Siddhānta is 266°

40′, which is a good match with the relative longitude of 265° 18′ in Aświnī-beginning Kṛttikā system.

9.23: The yogatārā of Śravaṇa

As shown in Figure 7.9 and Table 7.25, Śravaṇa has three stars in its star group with Altair (α Aql) being the accepted yogatārā. Altair is the brightest star in the group with apparent magnitude of 0.75 and J2000.0 ecliptic longitude and latitude of 301° 47′ and 29° 18′ respectively. As seen in Table 5.1, the longitude given in Sūrya Siddhānta is 280° 0′, which is a reasonable match with the relative longitude of 281° 33′ in Aświnī-beginning Kṛttikā system.

9.24: The yogatārā of Dhaniṣṭhā

As shown in Figure 7.9 and Table 7.26, Dhaniṣṭhā has four stars in its star group with Rotanev (β Del) being the accepted yogatārā with apparent magnitude of 4.10 and J2000.0 ecliptic longitude and latitude of 316° 20′ and 31° 55′ respectively. All stars in the group are dim stars with apparent magnitude varying from 3.85 to 4.40. As seen in Table 5.1, the longitude given in Sūrya Siddhānta is 290° 0′, which is 6° 23′ lesser than the relative longitude of 296° 23′ of Rotanev (β Del) in Aświnī-beginning Kṛttikā system. Latitude given in Sūrya Siddhānta is 36° 0′, which is 4° 5′ greater than the J2000.0 ecliptic latitude of 31° 55′ of Rotanev (β Del).

In this case, there is a star in the group that fits the longitude and latitude given in Sūrya Siddhānta better than Rotanev (β Del). The star Al Salib (γ2 Del) has apparent magnitude of 4.25 and J2000.0 ecliptic longitude and latitude of 319° 22′ and 32° 42′ respectively. It has relative longitude of 289° 57′ in Aświnī-beginning Rohiṇī system, which is an excellent match with the 290° 0′ longitude given in Sūrya Siddhānta. Since Al Salib (γ2 Del) is a better fit in terms of both longitude and latitude compared to Rotanev (β Del), Al Salib (γ2 Del) has a better claim of being the yogatārā of Dhaniṣṭhā. Figure 9.5 shows the proposed and accepted yogatārās of Dhaniṣṭhā nakṣatra.

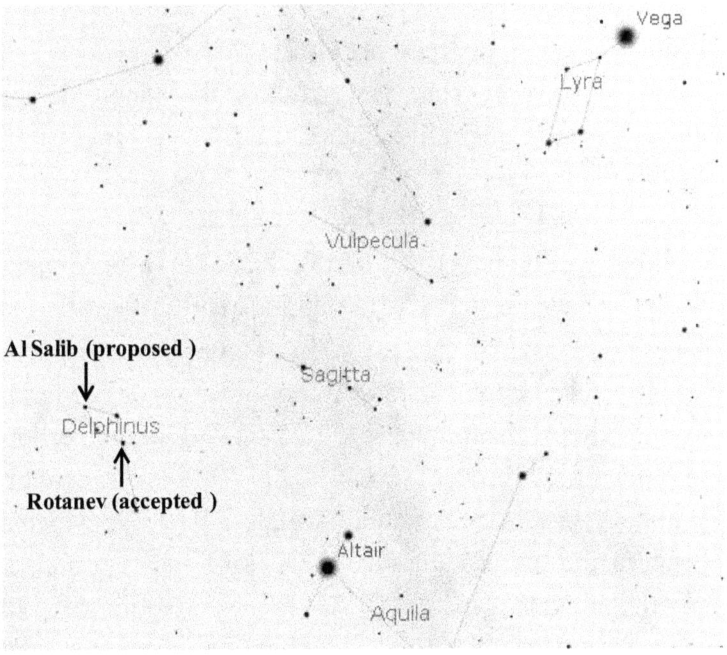

Figure 9.5: The proposed and accepted yogatārās of Dhaniṣṭhā nakṣatra.

9.25: The yogatārā of Śatabhiṣaja

As shown in Figure 7.10 and Table 7.27, Śatabhiṣaja has only one star in its star group, Hydor (λ Aqr), which by default is its yogatārā. Śatabhiṣaja has apparent magnitude of 3.70 and J2000.0 ecliptic longitude and latitude of 341° 35′ and -0° 23′ respectively. As seen in Table 5.1, the longitude given in Sūrya Siddhānta is 320° 0′, which is a reasonable match with the relative longitude of 321° 34′ in Aświnī-beginning Kṛttikā system. The latitude given in Sūrya Siddhānta is -0° 30′, which means the star is very close to ecliptic. There is no visibly bright star very close to ecliptic within 5 degrees on either side of Hydor. Thus Hydor is the only star that matches the longitude and latitude given in Sūrya Siddhānta.

9.26: The yogatārā of Pūrva-bhādrapadā

As shown in Figure 7.10 and Table 7.28, Pūrva-bhādrapadā has two stars in its star group currently identified as Markab (α Peg) and Scheat (β Peg). Out of these two stars, currently accepted yogatārā of Pūrva-bhādrapadā is Markab with apparent magnitude of 2.45 and J2000.0 ecliptic longitude and latitude of 353° 29′ and 19° 24′ respectively. Latitude given in Sūrya Siddhānta is 24° 0′ N, which is closer to the average of the latitudes of the two stars in Pūrva-bhādrapadā nakṣatra. As seen in Table 5.1, the longitude given in Sūrya Siddhānta is 326° 0′, which matches reasonably with the longitude of 323° 57′ in Aświnī-beginning Rohiṇī system.

9.27: The yogatārā of Uttara-bhādrapadā

As shown in Figure 7.10 and Table 7.29, Uttara-bhādrapadā has two stars in its star group currently identified as Algenib (γ Peg) and Alpheratz (α And/δ Peg). Latitude given in Sūrya Siddhānta is 26° 0′ N, which matches well with the J2000.0 ecliptic latitude of 25° 41′ of Alpheratz. As pointed out by Burgess [1], the latitude given in Sūrya Siddhānta matches that of Alpheratz (α And/δ Peg), while the longitude matches that of Algenib (γ Peg), the other star in the group. The longitude of Algenib is 339° 36′ in Aświnī-beginning Rohiṇī system, which matches reasonably with the longitude of 337° 0′ given in Sūrya Siddhānta. The identity of the stars in the Pūrva-bhādrapadā and Uttara-bhādrapadā nakṣatras is beyond doubt due to the four stars belonging to these two nakṣatras forming a highly visible near square shape in the sky.

9.28: The yogatārā of Revatī

As shown in Figure 7.11 and Table 7.30, Revatī has only one stars in its star group, Revati (ζ Psc A), which by default is its yogatārā. Revati is a very dim star with apparent magnitude of 5.20 and J2000.0 ecliptic longitude and latitude of 19° 53′ and -0° 13′ respectively. As seen in Table 5.1, the longitude given in Sūrya Siddhānta is 359° 50′, which has been used to determine the coordinates of other yogatārās in the Aświnī-beginning Kṛttikā system. As discussed before, this longitude of 359° 50′ is an excellent match with the longitude of 319° 44′ of Revati (ζ Psc A) in the Kṛttikā system as the boundaries of Revatī and Kṛttikā nakṣatras are separated by 40°.

Based on the reassessment of the identifications of yogatārās above, alternative identifications of five yogatārās have been proposed. Table 9.4 shows the J2000.0 ecliptic coordinates of these five yogatārās along with the longitudes given in Sūrya Siddhānta. The ecliptic coordinates of other yogatārās have been listed in Table 7.1.

Table 9.4: Alternate Yogatārā identifications

No.	Nakṣatra	Junction-star (yogatārā)	Ecliptic Longitude*	Ecliptic Latitude*
1	Aświnī	Sheratan (β Ari)**	33° 58′	8° 29′ N
1	*Aświnī*	*Hamal (α Ari)****	*37° 40′*	*9° 58′*
2	Bharaṇī	Barani II (35 Ari)**	46° 56′	11° 19′ N
2	*Bharaṇī*	*Bharani (41 Ari)****	*48° 12′*	*10° 27′ N*
13	Hasta	Algorab (δ Crv)**	193° 27′	12° 12′ S
13	*Hasta*	*Gienah (γ Crv)****	*190° 44′*	*14° 30′ S*
15	Swāti	Arcturus (α Boo)**	204° 14′	30° 44′ N
15	*Swāti*	*Alphecca (α CrB)****	*222° 18′*	*44° 19′ N*
24	Dhaniṣṭhā	Rotanev (β Del)**	316° 20′	31° 55′ N
24	*Dhaniṣṭhā*	*Al Salib (γ2 Del)****	*319° 22′*	*32° 44′ N*

* J2000.0 ecliptic coordinates based on Stellarium software.
** Identifications by Burgess [1].
*** This study.

As pointed out above, the longitudes given in Sūrya Siddhānta have some coordinates from Aświnī-beginning Rohiṇī system, some coordinates from Aświnī-beginning Kṛttikā system, and some coordinates resulting from corrections made due to confusion between Rohiṇī and Kṛttikā systems. A comparison of longitudes of yogatārās in these different systems is presented in Table 9.5 to show the best fit with the longitudes given in Sūrya Siddhānta.

Table 9.5: Comparison of longitudes of yogatārās

	1. Aświnī	2. Bharaṇī	3. Kṛttikā	4. Rohiṇī
	α Ari	41 Ari	η Tau	α Tau
Sūrya Siddhānta	8° 0′	20° 0′	37° 30′	49° 30′
Aświnī-beginning Rohiṇī system	8° 0′	18° 36′	30° 23′	40° 09′
Aświnī-beginning Kṛttikā offset	14° 18′	24° 53′	36° 40′	46° 26′
Aświnī-beginning Kṛttikā system	17° 38′	28° 13′	40° 0′	49° 46′
Best fit	Aświnī-beginning Rohiṇī system	Aświnī-beginning Rohiṇī system	Aświnī-beginning Kṛttikā offset	Aświnī-beginning Kṛttikā system
Deviation	0° 0′	-1° 24′	-0° 50′	0° 16′

Table 9.5: Comparison of longitudes of yogatārās (continued)

	5. Mṛgaśirā	6. Ārdrā	7. Punarvasu	8. Puṣya
	λ Ori	α Ori	β Gem	δ Cnc
Sūrya Siddhānta	63° 0′	67° 20′	93° 0′	106° 0′
Aświnī-beginning Rohiṇī system	54° 05′	59° 07′	83° 57′	99° 05′
Aświnī-beginning Kṛttikā offset	60° 22′	65° 25′	90° 7′	105° 22′
Aświnī-beginning Kṛttikā system	63° 42′	68° 45′	93° 27′	108° 42′
Best fit	Aświnī-beginning Kṛttikā system	Aświnī-beginning Kṛttikā system	Aświnī-beginning Kṛttikā system	Aświnī-beginning Kṛttikā offset

Deviation	0° 42′	1° 25′	0° 27′	-0° 38′

Table 9.5: Comparison of longitudes of yogatārās (continued)

	9. Āśleṣā	10. Maghā	11. Pūrva-phālgunī	12. Uttara-phālgunī
	ε Hya	α Leo	δ Leo	β Leo
Sūrya Siddhānta	109° 0′	129° 0′	144° 0′	155° 0′
Aświnī-beginning Rohiṇī system	102° 55′	120° 22′	131° 31′	142° 11′
Aświnī-beginning Kṛttikā offset	109° 8′	126° 36′	137° 52′	148° 25′
Aświnī-beginning Kṛttikā system	112° 28′	129° 56′	141° 12′	151° 45′
Best fit	Aświnī-beginning Kṛttikā offset	Aświnī-beginning Kṛttikā system	Aświnī-beginning Kṛttikā system	Aświnī-beginning Kṛttikā system
Deviation	0° 8′	0° 56′	-2° 48′	-3° 15′

Table 9.5: Comparison of longitudes of yogatārās (continued)

	13. Hasta	14. Citrā	15. Swāti	16. Viśākhā
	γ Crv	α Vir	α CrB	ι Lib
Sūrya Siddhānta	170° 0′	180° 0′	199° 0′	213° 0′
Aświnī-beginning Rohiṇī system	161° 17′	174° 15′	192° 22′	201° 25′
Aświnī-beginning Kṛttikā offset	167° 30′	180° 32′	198° 45′	207° 41′
Aświnī-beginning Kṛttikā system	170° 50′	183° 52′	202° 05′	211° 01′
Best fit	Aświnī-beginning Kṛttikā system	Aświnī-beginning Kṛttikā offset	Aświnī-beginning Kṛttikā offset	Aświnī-beginning Kṛttikā system
Deviation	0° 50′	0° 32′	-0° 15′	-1° 59′

Table 9.5: Comparison of longitudes of yogatārās (continued)

	17. Anurādhā	18. Jyeṣṭhā	19. Mūla	20. Pūrvāṣāḍhā
	δ Sco	α Sco	λ Sco	δ Sgr
Sūrya Siddhānta	224° 0′	229° 0′	241° 0′	254° 0′
Aświnī-beginning Rohiṇī system	212° 58′	220° 10′	234° 59′	244° 57′
Aświnī-beginning Kṛttikā offset	219° 15′	226° 26′	241° 15′	251° 14′
Aświnī-beginning Kṛttikā system	222° 35′	229° 46′	244° 35′	254° 34′
Best fit	Aświnī-beginning Kṛttikā system	Aświnī-beginning Kṛttikā system	Aświnī-beginning Kṛttikā offset	Aświnī-beginning Kṛttikā system
Deviation	-1° 25′	0° 46′	0° 15′	0° 34′

Table 9.5: Comparison of longitudes of yogatārās (continued)

	21. Uttarāṣāḍhā	22. Abhijit	23. Śravaṇa	24. Dhaniṣṭhā
	σ Sgr	α Lyr	α Aql	γ2 Del
Sūrya Siddhānta	260° 0′	266° 40′	280° 0′	290° 0′
Aświnī-beginning Rohiṇī system	252° 46′	255° 40′	271° 50′	289° 57′
Aświnī-beginning Kṛttikā offset	259° 3′	261° 58′	278° 13′	296° 11′
Aświnī-beginning Kṛttikā system	262° 23′	265° 18′	281° 33′	299° 31′
Best fit	Aświnī-beginning Kṛttikā offset	Aświnī-beginning Kṛttikā system	Aświnī-beginning Kṛttikā system	Aświnī-beginning Rohiṇī system
Deviation	-0° 57′	-1° 22′	1° 33′	-0° 3′

Table 9.5: Comparison of longitudes of yogatārās (continued)

	25. Śatabhiṣaja	26. Pūrva-bhādrapadā	27. Uttara-bhādrapadā	28. Revatī
	λ Aqr	α Peg	γ Peg	ζ Psc
Sūrya Siddhānta	320° 0′	326° 0′	337° 0′	359° 50′
Aświnī-beginning Rohiṇī system	311° 57′	323° 57′	339° 36′	350° 12′
Aświnī-beginning Kṛttikā offset	318° 14′	330° 12′	345° 52′	356° 30′
Aświnī-beginning Kṛttikā system	321° 34′	333° 32′	349° 12′	359° 50′
Best fit	Aświnī-beginning Kṛttikā system	Aświnī-beginning Rohiṇī system	Aświnī-beginning Rohiṇī system	Aświnī-beginning Kṛttikā system
Deviation	1° 34′	-2° 3′	2° 36′	0° 0′

Some researchers have used the best fit method to date Sūrya Siddhānta. Burgess [1] had assumed a base year of 560 CE for identifications of yogatārās. He then calculated the average error in longitudes of the yogatārās and came to the conclusion that the star coordinates given in Sūrya Siddhānta were measured around 490 CE. Abhyankar [3] used a least square method to conclude that the best fit was obtained for 430 CE. Pingree and Morrissey [2] compared the star coordinates given in Sūrya Siddhānta with star coordinates in 400, 425, 450, 475 and 500 CE and concluded that best fit was close to 425 CE. As discussed above, the longitudes given in Sūrya Siddhānta are inconsistent due to a mix up between different systems and therefore a best fit approach cannot be applied to determine the date of observations. The reason for this mix up seems to be related to the reinstatement of Revatī nakṣatra in the list of nakṣatras. Atharvaveda Saṃhitā (19.7.1-5) describes a system of 28 nakṣatras. Later texts adopted a system of 27 nakṣatras by dropping Abhijit from the list. However, there was also an alternate system in which Revatī nakṣatra was dropped from the list to make way for 27 nakṣatra system. The story of dropping of Revatī and her reinstatement is described in Chapter 72 of Mārkaṇḍeya Puarāṇa. This could also explain why the yogatārā of Revatī is so dim. Currently accepted yogatārā of Revatī was not the original yogatārā of Revatī. Jain text Jambudwīpaprajñapti 7.191 says that the Revatī nakṣatra has 32 stars and Jambudwīpaprajñapti 7.189 says that their positions are north of the trajectory of moon. The yogatārā of Revatī nakṣatra is currently identified as Revati (Kuton II), which is close to ecliptic and hence not north of the trajectory of moon. The reinstatement of Revatī nakṣatra may have resulted in the choice of a very dim star as the yogatārā of Revatī nakṣatra and shifting of the ecliptic longitude of the other yogatārās due to the shift of the zero point for measurement of longitude.

Misidentification of the yogatārās has been investigated by Abhyankar [3] and Venkatachar [4]. Abhyankar [3] identified the yogatārās of Bharaṇī as Bharani (41 Ari) instead of Barani II (35 Ari), Ārdrā as Alhena (γ Gem) instead of Betelguese (α Ori), Āśleṣā as Minazal V (ζ Hya) instead of Minazal III (ε Hya), Hasta as Gienah (γ Crv) instead of Algorab (δ Crv), Viśākhā as Zubenelgenubi (α Lib) instead of HIP 74392 (ι Lib), Abhijit as Altair (α Aql) instead of Vega (α Lyr), Śravaṇa as Rotanev (β Del) instead of Altair (α Aql), Dhaniṣṭhā as Sadalsuud (β Aqr) instead of Rotanev (β Del), and Śatabhiṣaja as Fomalhaut (α PsA) instead of Hydor (λ Aqr). Out of these identifications, Bharani (41 Ari) for Bharaṇī and Gienah (γ Crv) for Hasta are same as this study. Abhyankar [3] has justified the identifications based on the accepted yogatārās not being within their boundaries or being too far away from ecliptic. Both of these reasons are unjustified. Many yogatārās are outside their boundaries in uniform division of ecliptic because the original span of nakṣatras was not uniform. There was no constraint on yogatārās to be close to ecliptic, which is evident from the latitudes of many yogatārās being over 20 degrees. Also, many of these identifications completely disregard the latitudes given in Sūrya Siddhānta, and hence are unacceptable.

Venkatachar [4] has proposed the yogatārās of Aświnī as Mirach (β And) instead of Sheratan (β Ari), Bharaṇī as Hamal (α Ari) instead of Barani II (35 Ari), Mṛgaśirā as Betelguese (α Ori) instead of Meissa (λ Ori), Ārdrā as Alhena (γ Gem) instead of Betelguese (α Ori), Puṣya as Al Tarf (β Cnc) instead of Asellus Australis (δ Cnc), Āśleṣā as Algenubi (ε Leo) instead of Minazal III (ε Hya), Hasta as Gienah (γ Crv) instead of Algorab (δ Crv), Swāti as Alphecca (α CrB) instead of Arcturus (α Boo), Viśākhā as Zubenelgenubi (α Lib) instead of HIP 74392 (ι Lib), Jyeṣṭhā as Hip 82396 (ε Sco) instead of Antares (α Sco), Pūrvāṣāḍhā as Nunki (σ Sgr) instead of Kaus Media (δ Sgr), Uttarāṣāḍhā as Altair (α Aql) instead of Nunki (σ Sgr), Śravaṇa as Rotanev (β Del) instead of Altair (α Aql), Dhaniṣṭhā as Sadalsuud (β Aqr) instead of Rotanev (β Del), Śatabhiṣaja as Fomalhaut (α PsA) instead of Hydor (λ

Aqr), and Revatī as Alpheratz (α And) instead of Revati (ζ Psc A). Out of these identifications, Gienah (γ Crv) for Hasta and Alphecca (α CrB) for Swāti are same as this study. Venkatachar [4] has justified the identifications based on the accepted yogatārās not being within their boundaries. This reasoning, as discussed above, is unjustified. Similar to Abhyankar [3], many of these identifications completely disregard the latitudes given in Sūrya Siddhānta, and hence are unacceptable.

As discussed in Chapter 1, currently accepted version of Indian history is based on the wrong identifications of the sheet anchors of Indian history. Since this distorted history needs to match the astronomical observations, zero points of Vedic astronomy are also shifted by six centuries. With the rediscovery of coordinate systems used by Indian astronomers, the correct zero points of Vedic astronomy can be established.

Notes

1. Burgess (1860).
2. Pingree and Morrissey (1989).
3. Abhyankar (1991).
4. Venkatachar (2014).

"This star constellation will be our guide and point of reference from the beginning to the end."
– Andrea Vogel

10. Zero Points of Vedic Astronomy

The zero point of ecliptic longitude is called the First Point of Aries, which is the position of vernal equinox on the ecliptic. As this point keeps changing due to precession, ecliptic longitudes are specified with the associated year. Currently, it is customary to specify the ecliptic coordinates for year 2000 called J2000 coordinates. Earlier, the ecliptic coordinates were specified for year 1950 called B1950 coordinates, and before that for year 1900 called J1900 coordinates. Due to precession, the First Point of Aries does not fall in the constellation of Aries anymore. The difference in angle between the First Point of Aries and the beginning of Aries constellation is the ayanāṃśa for western astronomy and astrology.

For ancient Indian astronomy and Vedic sidereal astrology, ayanāṃśa is the difference in angle between the current position of vernal equinox and zero point of nakṣatra system. There is no unanimity regarding the zero point of nakṣatra system, and therefore there are many values of ayanāṃśa currently in use. The most popular ayanāṃśa value is Lahiri ayanāṃśa, which was recommended by the

Calendar Reform Committee set up by the Government of India. The Calendar Reform Committee was chaired by Prof. M. N. Saha. Mr. N. C. Lahiri acted as the Secretary of the Committee. It gave its report in 1955, in which following is said about the zero point of the Hindu Zodiac:

> The zero point of the Hindu Zodiac: By this is meant the Vernal Equinoctial point (first point of Aries) at the time when the Hindu savants switched on from the old Vedāṅga-Jyotiṣa calendar to the Siddhāntic calendar (let us call this epoch of the Siddhānta-Jyotiṣa or S.J.). There is a wide spread belief that a definite location can be found for this point from the data given in the Sūrya-Siddhānta and other standard treatises. This impression is however wrong.
>
> Its location has to be inferred from the co-ordinates given for known stars in Chap. VIII of the Sūrya-Siddhānta. From these data Dīkṣit thought that he had proved that it was very close to Revatī (ζ Piscium); but another school thinks that the autumnal equinoctial point (first point of Libra) at this epoch was very close to the star Citrā (Spica, α Virginis), and therefore the first point of Aries at the epoch of S.J. was 180° behind this point. The celestial longitude in 1950 of ζ Piscium was 19° 10′39″ and of α Virginis was 203° 8′36″. The longitudes of the first point of Aries, according to the two schools therefore differ by 23° 9′ (-) 19° 11′ = 3° 58′ and they cannot be identical. Revatī or ζ Piscium was closest to Υ_0 (the V.E. point) about 575 A.D., and Citrā or α Virginis was closest to $\underline{\Omega}$ (the A.E. point) about 285 A.D., a clear difference of 290 years.
>
> Thus even those who uphold the nirāyaṇa school are not agreed amongst themselves regarding the exact location

> of the vernal point in the age of the Sūrya-Siddhānta and though they talk of the Hindu zero-point, they do not know where it is. [1]

As discussed in previous two chapters, the exact zero points of Vedic-Hindu astronomy can be determined from a detailed analysis of coordinates given in Sūrya-Siddhānta. Indian astronomers had developed two different co-ordinate systems, Rohiṇī system with the origin at Aldebaran, the yogatārā of Rohiṇī nakṣatra, and Kṛttikā system with the origin at Alcyone, the yogatārā of Kṛttikā nakṣatra.

10.1 Rohiṇī system

The Rohiṇī system, as illustrated in Figure 8.1, has the beginning of Rohiṇī nakṣatra at Aldebaran and nakṣatra boundaries at 13° 20′ intervals from Aldebaran. Zero points of Vedic astronomy in Rohiṇī system correspond to the position of vernal equinox at nakṣatra boundaries in Rohiṇī system and are shown in Table 10.1. Duration of vernal equinox in each nakṣatra from Rohiṇī to Pūrva-bhādrapadā is shown in Figure 10.1 and Table 10.2.

10.2 Kṛttikā system

The Kṛttikā system, as illustrated in Figure 8.2, has the end of Kṛttikā nakṣatra at Alcyone and nakṣatra boundaries at 13° 20′ intervals from Alcyone. Zero points of Vedic astronomy in Kṛttikā system correspond to the position of vernal equinox at nakṣatra boundaries in Kṛttikā system and are shown in Table 10.3. Duration of vernal equinox in each nakṣatra in Kṛttikā system from Kṛttikā to Pūrva-bhādrapadā is shown in Figure 10.2 and Table 10.4.

Table 10.1: Zero points of Vedic astronomy in Rohiṇī system

Boundary	Star	Longitude	Year
Rohiṇī/Kṛttikā	Aldebaran	0° 0′	3045 BCE
Kṛttikā/Bharaṇī	Aldebaran	13° 20′	2074 BCE
Bharaṇī/Aświnī	Aldebaran	26° 40′	1106 BCE
Aświnī/Revatī	Aldebaran	40° 0′	142 BCE
Revatī/ Uttara-bhādrapadā	Aldebaran	53° 20′	820 CE
Uttara-bhādrapadā/Pūrva-bhādrapadā	Aldebaran	66° 40′	1776 CE
Pūrva-bhādrapadā/ Śatabhiṣaja	Aldebaran	80° 0′	2729 CE

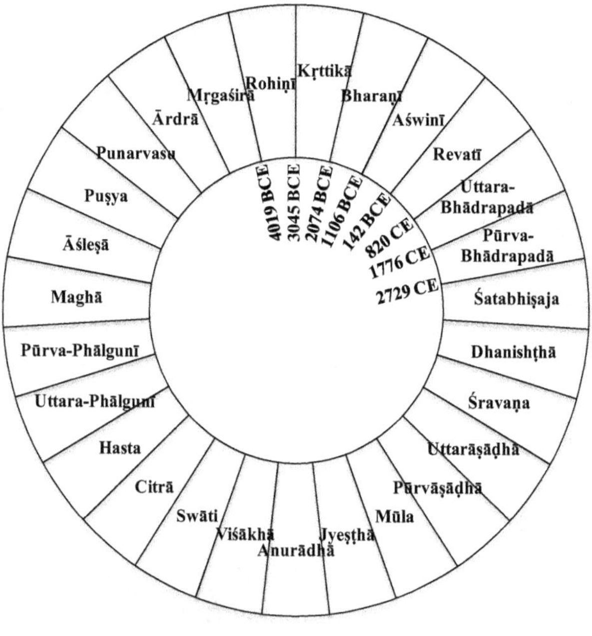

Figure 10.1: Position of vernal equinox in Rohiṇī system

Table 10.2: Duration of vernal equinox in Rohiṇī system

Nakṣatra	Duration
Rohiṇī	4019 BCE to 3045 BCE
Kṛttikā	3045 BCE to 2074 BCE
Bharaṇī	2074 BCE to 1106 BCE
Aświnī	1106 BCE to 142 BCE
Revatī	142 BCE to 820 CE
Uttara-bhādrapadā	820 CE to 1776 CE
Pūrva-bhādrapadā	1776 CE to 2729 CE

Table 10.3: Zero points of Vedic astronomy in Kṛttikā system

Boundary	Star	Longitude	Year
Rohiṇī/Kṛttikā	Alcyone	0° 0′	2337 BCE
Kṛttikā/Bharaṇī	Alcyone	13° 20′	1367 BCE
Bharaṇī/Aświnī	Alcyone	26° 40′	400 BCE
Aświnī/Revatī	Alcyone	40° 0′	563 CE
Revatī/ Uttara-bhādrapadā	Alcyone	53° 20′	1522 CE
Uttara-bhādrapadā/Pūrva-bhādrapadā	Alcyone	66° 40′	2477 CE
Pūrva-bhādrapadā/ Śatabhiṣaja	Alcyone	80° 0′	3428 CE

Table 10.4: Duration of vernal equinox in Kṛttikā system

Nakṣatra	Duration
Kṛttikā	2337 BCE to 1367 BCE
Bharaṇī	1367 BCE to 400 BCE
Aświnī	400 BCE to 563 CE
Revatī	563 CE to 1522 CE
Uttara-bhādrapadā	1522 CE to 2477 CE
Pūrva-bhādrapadā	2477 CE to 3428 CE

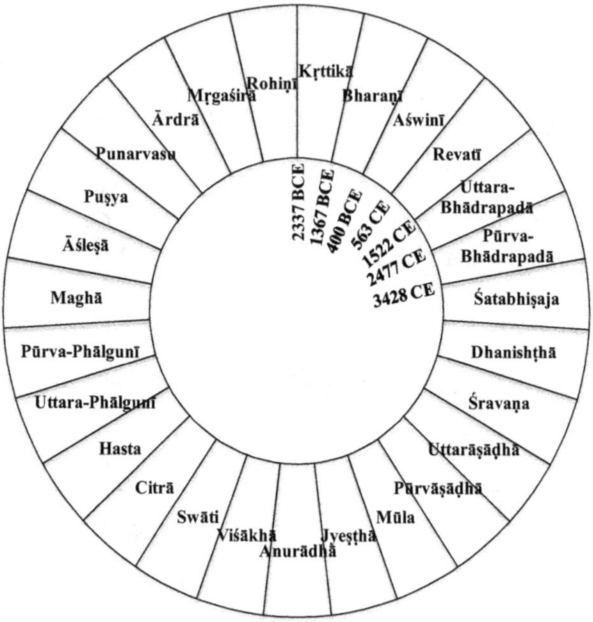

Figure 10.2: Position of vernal equinox in Kṛttikā system

Once the zero points of Vedic astronomy have been determined, it is straightforward to calculate the ayanāṃśa values corresponding to these zero points.

Notes

1. CSIR (1955): 262

"Everything that is beautiful and noble is the product of reason and calculation."
- Charles Baudelaire

11. The Calculation of Ayanāṃśa

The Calendar Reform Committee set up by the Government of India gave its report in 1955 and recommended that the value of ayanāṃśa be taken as 23° 15′ on 21st March, 1956 [1]. This corresponds to autumnal equinox falling on Spica or α Virginis, the yogatārā of Citrā, which the Calendar Reform Committee calculated as having taken place in 285 CE. It is equivalent to the zero point of Hindu Zodiac being at the vernal equinoctial point at 180° from Spica. The value of ayanāṃśa is closer to 24° now. The choice of zero point adopted by the Calendar Reform Committee is called Citrā-Pakṣa. The choice of zero point as the vernal equinoctial point falling on Revati or ζ Piscium, the yogatārā of Revatī, is called Revatī-Pakṣa. It should be noted that the Calendar Reform Committee did not provide any scientific reasoning for choosing Citrā-Pakṣa over Revatī-Pakṣa. Since there is no clarity regarding the location of the zero point of the Hindu zodiac, there are many ayanāṃśa values currently in use besides the Lahiri ayanāṃśa based on Citrā-Pakṣa.

Ayanāṃśa was never supposed to be more than 13° 20′. As vernal equinox crossed the boundary between nakṣatras, the order of the nakṣatras was changed to reflect the new position of vernal equinox. As discussed earlier, two ecliptic coordinate systems were used in India, one with the yogatārā of Rohiṇī, Aldebaran, at zero point and the other with the yogatārā of Kṛttikā, Alcyone, at zero point. The values of Ayanāṃśa in these two systems are described below.

11.1 Rohiṇī system

The Rohiṇī system has the beginning of Rohiṇī nakṣatra at Aldebaran and nakṣatra boundaries at 13° 20′ intervals from Aldebaran. Ayanāṃśa values at zero points in the Rohiṇī system are shown in Table 11.1 and correspond to the position of vernal equinox at nakṣatra boundaries in Rohiṇī system as shown in Figure 11.1.

Ayanāṃśa values in 2000 CE in Rohiṇī system were obtained by setting the date to January 1, 2000 at noon in Stellarium. The ayanāṃśa at Rohiṇī/Kṛttikā boundary was obtained by noting the J2000 longitude of Aldeberan, the yogatārā of Rohiṇī. Since Aldeberan is at the zero point of this system, J2000 longitude of Aldeberan represents how much the vernal equinox had moved from Aldeberan in year 2000. Ayanāṃśa values in 2000 CE at zero points at Kṛttikā/Bharaṇī, Bharaṇī/Aświnī, and Aświnī/Revatī boundaries were obtained by deducting 13° 20′, 26° 40′, and 40° 0′ respectively from the ayanāṃśa values at Rohiṇī/Kṛttikā boundary.

Table 11.1: Ayanāṃśa values at zero points in the Rohiṇī system

	Boundary	Star	Longitude	Ayanāṃśa in 2000 CE
1.	Rohiṇī/Kṛttikā	Aldebaran	0° 0′	69° 47′
2.	Kṛttikā/Bharaṇī	Aldebaran	13° 20′	56° 27′
3.	Bharaṇī/Aświnī	Aldebaran	26° 40′	43° 07′
4.	Aświnī/Revatī	Aldebaran	40° 0′	29° 47′

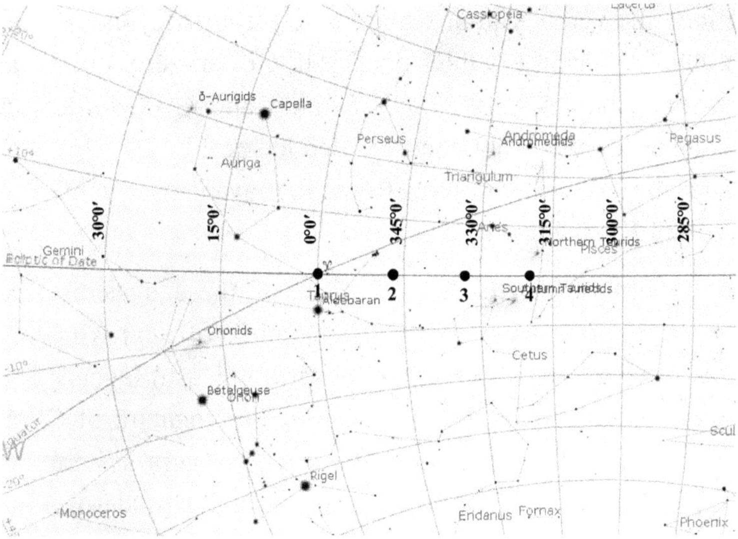

Figure 11.1: Zero points in the Rohiṇī system

11.2 Kṛttikā system

As discussed in Chapter 8, the yogatārā of Kṛttikā, Alcyone, is placed at the boundary between Rohiṇī and Kṛttikā in Kṛttikā system. The nakṣatra boundaries are at 13° 20′ intervals from Alcyone. Ayanāṃśa values at zero points in the Kṛttikā system are shown in Table 11.2 and correspond to the position of vernal equinox at nakṣatra boundaries in Kṛttikā system as shown in Figure 11.2.

Ayanāṃśa values in 2000 CE in Kṛttikā system were obtained by setting the date to January 1, 2000 at noon in Stellarium. The ayanāṃśa at Rohiṇī/Kṛttikā boundary was obtained by noting the J2000 longitude of Alcyone, the yogatārā of Kṛttikā. Since Alcyone is at the zero point of this system, J2000 longitude of Alcyone represents how much the vernal equinox had moved from Alcyone in year 2000. Ayanāṃśa values in 2000 CE at zero points at Kṛttikā/Bharaṇī, Bharaṇī/Aświnī, and Aświnī/Revatī boundaries were obtained by deducting 13° 20′, 26° 40′, and 40° 0′ respectively from the ayanāṃśa values at Rohiṇī/Kṛttikā boundary.

It should be noted that in none of these systems, the ayanāṃśa is closer to Lahiri ayanāṃśa of approximately 24° used by most calendar makers and astrologers. The Lahiri ayanāṃśa is based on Spica, the yogatārā of Citrā, being at 180° from the zero point of nakṣatra system. As shown in Tables 9.2 and 9.3, Spica has longitude of 174° 15′ in Rohiṇī system and 183° 52′ in Kṛttikā system. It is not diametrically opposite to the zero point in either system. The Lahiri ayanāṃśa is calculated from 285 CE, when Spica was at autumnal equinox, and it is assumed that vernal equinox was at zero point located diametrically opposite to Spica. However, as shown in Figures 10.1 and 10.2, the vernal equinox did not cross the boundary of any nakṣatra during third century in either Rohiṇī system or Kṛttikā system.

Table 11.2: Ayanāṃśa values at zero points in the Kṛttikā system

	Boundary	Star	Longitude	Ayanāṃśa in 2000 CE
1.	Rohiṇī/Kṛttikā	Alcyone	0° 0′	60° 00′
2.	Kṛttikā/Bharaṇī	Alcyone	13° 20′	46° 40′
3.	Bharaṇī/Aświnī	Alcyone	26° 40′	33° 20′
4.	Aświnī/Revatī	Alcyone	40° 0′	20° 00′

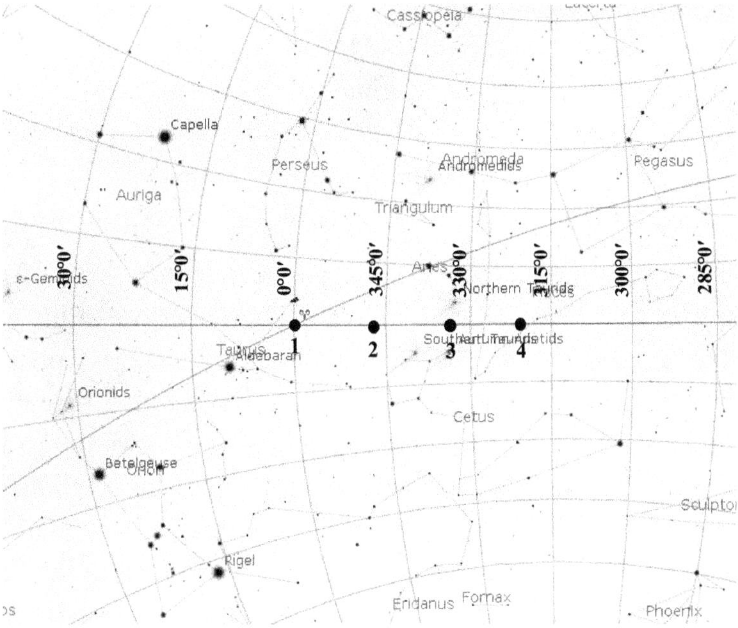

Figure 11.2: Zero points in the Kṛttikā system

As pointed earlier in this chapter, ayanāṃśa was never supposed to be more than 13° 20′. The order of nakṣatras was supposed to be updated as the vernal equinox crossed the boundaries of nakṣatras. This has not been done due to the loss of knowledge about the coordinate systems used by ancient Indian astronomers. The value of ayanāṃśa depends on the first nakṣatra in the nakṣatra list, which was last updated to begin with Aświnī. The value of ayanāṃśa from the beginning of Aświnī naksatra in Rohiṇī system was 29° 47′ in 2000 CE as shown in Table 11.1. The value of absolute ayanāṃśa was 69° 47′ in 2000 CE with Aldeberan at the zero point of the Rohiṇī system. The value of ayanāṃśa from the beginning of Aświnī naksatra in Kṛttikā system was 20° 00′ in 2000 CE as shown in Table 11.2. The value of absolute ayanāṃśa was 60° 00′ in 2000 CE with Alcyone at the zero point of the Kṛttikā system.

The understanding of the correct zero points of Vedic astronomy is crucial in discovering the true history of India. The earliest astronomical text of India is Vedāṅga Jyotiśa, and its date of composition has been forwarded in time by up to six and half centuries to match the forwarding of Maurya and Imperial Gupta dynasties by six and half centuries. A detailed analysis of the astronomical information is needed to understand why the current dating of the Vedāṅga Jyotiśa is possibly wrong.

Notes

1. CSIR (1955): 7

"Astronomy is, not without reason, regarded, by mankind, as the sublimest of the natural sciences. Its objects so frequently visible, and therefore familiar, being always remote and inaccessible, do not lose their dignity."

— Benjamin Silliman

12. The Dating of Vedāṅga Jyotiśa

Currently, the Vedāṅga Jyotiśa is dated between 1150 BCE to 1400 BCE [1]. The Vedāṅga Jyotiśa is the first astronomical text of ancient Indian civilization and the determination of the correct date of the composition of the Vedāṅga Jyotiśa is of vital importance in discovering the correct chronological history of India.

There is a very specific observation in the Vedāṅga Jyotiśa that makes it straight forward to calculate the date of its composition. It is mentioned in the verses 6-8 of the Yajur-Vedāṅga Jyotiṣa that the winter solstice was at the beginning of the Śraviṣṭhā (Dhaniṣṭhā) nakṣatra and the summer solstice was at the midpoint of the Āśleṣā nakṣatra at the time of its composition. Figure 12.1 shows the position of the solstices in the background of the nakṣatras as mentioned in the Vedāṅga Jyotiṣa.

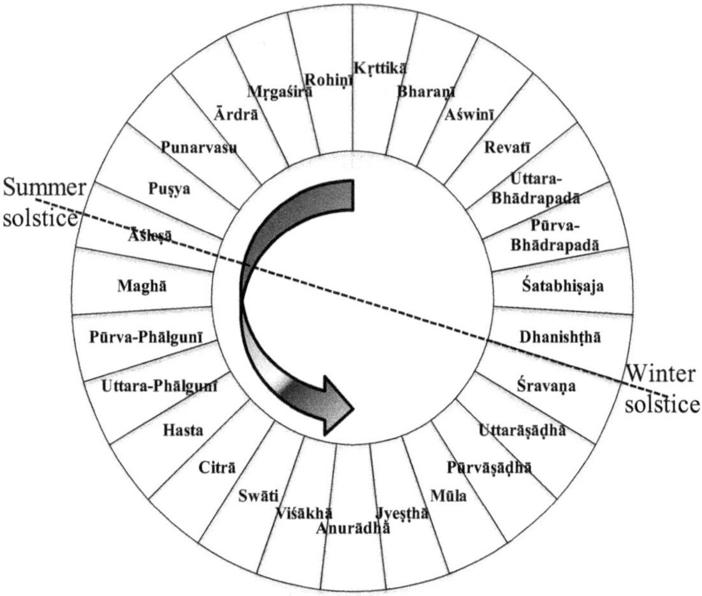

Figure 12.1: The position of solstices mentioned in Vedāṅga Jyotiṣa

Based on the information on the position of winter solstice in Vedāṅga Jyotiṣa, Kuppanna Sastry has calculated the date of the composition of the Vedāṅga Jyotiṣa using three different methods [1]. A critical analysis of each of these methods is presented below.

12.1 Method 1

Kuppanna Sastry has described the first method as follows:

> Verses 6, 7 and 8 of the Yajur-Vedāṅga Jyotiṣa show that at the time of Lagadha the winter solstice was at the

> beginning of the asterism Śraviṣṭhā (Delphini) segment and that the summer solstice was at the midpoint of the Āśleṣā segment. It can be seen that this is the same as alluded to by Varāhamihira in his Pañcasiddhāntikā and Bṛhatsaṃhitā. Since VM has stated that in his own time the summer solstice was at Punarvasu ¾, and the winter solstice at Uttarāṣāḍhā ¼, there had been a precession of 1 ¾ stellar segments, i.e. 23° 20'. From this we can compute that Lagadha's time was 72 x 23 1/3 = 1680 years earlier than VM's time (c. A.D. 530), i.e. c. 1150 B.C. If, instead of the segment, the group itself is meant, which is about 3° within it, Lagadha's time would be c. 1370 B.C. [1]

The date of Vedāṅga Jyotiṣa in this method is calculated by the difference in the position of solstices during the time of the composition of the Vedāṅga Jyotiṣa and the time of Varāhamihira. The time difference between these observations is roughly 1,680 years. The problem with this method is that it assumes the time of Varāhamihira to be circa 530 CE. However, this dating itself is a matter of contention. Indian tradition places Varāhamihira in 1st century BCE in the court of Emperor Vikramāditya along with many other luminaries including Kālidāsa. Currently accepted version of Indian history denies the historicity of Vikramāditya and places Varāhamihira and Kālidāsa in different time periods. Varāhamihira is placed in the sixth century based on assuming the zero point of Śaka era in 78 CE. However, it has been worked out by Venkatachelam that the Śaka era referred by Varāhamihira has the zero point in 550 BCE [2]. Thus the date of Vedāṅga Jyotiṣa arrived by this method remains doubtful.

12.2 Method 2

Kuppanna Sastry has described the second method as follows:

> The date arrived at as above can be confirmed by the Sūryasiddhānta and the Siddhānta Śiromaṇi which give 290° polar longitude and 36° polar latitude to Śraviṣṭhā. From this, the actual longitude of Śraviṣṭhā got is 296° 15'. Since the siddhāntas use the fixed zodiac beginning with the vernal equinox of c. 550 A.D., and the winter solstice of this is 270°, there has been a precession of 296° 15' - 270° = 26° 15'. Since 26 ¼ x 72 = 1890 years, the wanted time is 1890 years, before A.D. 550, i.e. c. 1340 B.C., being the same as the above, the small difference being observational. [1]

The date of Vedāṅga Jyotiṣa in this method is calculated by the difference in the position of the yogātarā of Śraviṣṭhā during the time of the composition of the Vedāṅga Jyotiṣa and Sūrya Siddhānta. This method assumes that the Sūrya Siddhānta uses the fixed zodiac beginning with the vernal equinox of circa 550 CE. As shown in previous two chapters, there were different zero points in Vedic-Hindu astronomy based on the position of the vernal equinox at nakṣatra boundaries. Most notably, the coordinates of many yogataras given in Sūrya Siddhānta match well with the Aświnī-beginning Rohiṇī system, which has its zero point in circa 142 BCE as shown in Table 10.1 and Figure 10.1. Thus the date of Vedāṅga Jyotiṣa arrived by this method remains doubtful. This brings us to the third method used by Kuppanna Sastry to calculate the date of Vedāṅga Jyotiṣa, which Kuppanna Sastry claims to be a direct method without depending on any other dating.

12.3 Method 3

Kuppanna Sastry has described the third method as follows:

> We can also calculate the time directly by comparing the position of Śraviṣṭhā (β Delphini) at the time when the winter solstice was 270°, with its position in 1940 A.D. (Rt. as. $20^h\ 36^m\ 51^s$ = 309° 13′, and declination 15° 42′ N). In the figure: The obliquity is about 23° 40′, γ is the vernal equinox, S is Śraviṣṭhā and R its Rt. As. position. Rγ = 360° – Rt. as. = 50° 47′. RS is the declination = 15° 42′. RS′ is the continuation of SR up to the ecliptic. Now:

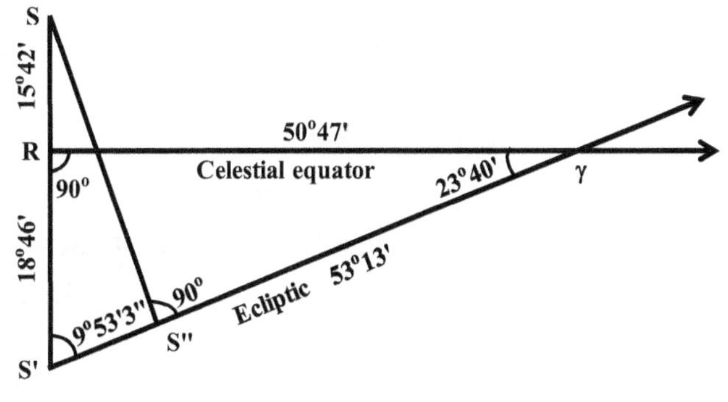

(i) From the rt. angled spherical triangle RγS′, it can be calculated that RS′ = 18° 46′; S′γ = 53°13′; and angle S′ = 75° 17′.

(ii) From the rt. angled spherical triangle SS′S″, S′S″ = 9° 53′3″, S″ being the celestial longitude of S in A.D. 1940. It was at 270° at the time required. Therefore, the precession is 360° – 53° 13′ – 270° + 9° 53′ = 46° 40′. Multiplying by 72, the time is 3360 years before A.D. 1940, i.e. c. 1400 B.C. If the beginning of the segment is

> meant and Śraviṣṭhā is about 3° inside, it is c. 1180 B.C. Since all these is subject to small errors of observation, it would be noted that we have got from all almost the same date for VJ. [1]

The date of Vedāṅga Jyotiṣa in this method is calculated by the difference in the position of the yogātarā of Śraviṣṭhā during the time of the composition of the Vedāṅga Jyotiṣa and 1940 CE. There is no need to get into the details of the mathematical calculation performed by Kuppanna Sastry, as the aim was simply to estimate the date when β Delphini was at winter solstice. Nowadays the date can be accurately determined using astronomical software such as Stellarium.

To obtain the date when β Delphini was at winter solstice using Stellarium, it is important to note that the ecliptic longitude of the winter solstice is 270° as shown in Figure 12.2. The date obtained by Stellarium is circa 1350 BCE as shown in Figure 12.3. Kuppanna Sastry estimates that β Delphini was at winter solstice circa 1400 BCE. This matches reasonably with the date obtained by Stellarium. Kuppanna Sastry next says that the currently accepted yogatara of Śraviṣṭhā, β Delphini, was 3° inside the Śraviṣṭhā nakṣatra and based on that the winter solstice was at the beginning of Śraviṣṭhā nakṣatra in circa 1180 BCE. Kuppanna Sastry then notes that all three methods yield reasonably matching dates between 1150 BCE to 1400 BCE, the difference being attributable to observational errors. It all seems fine until it is realized that the information given in Sūrya Siddhānta has been fudged by the interpretation of the given coordinates as polar coordinates.

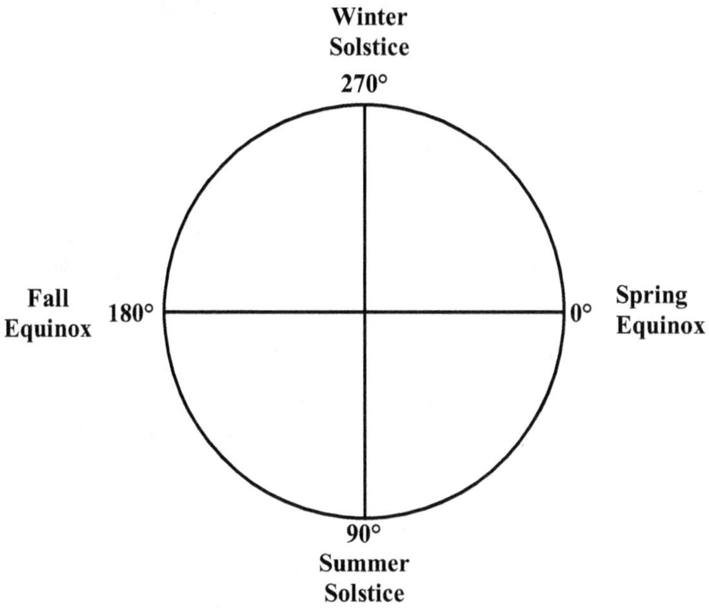

Figure 12.2: Ecliptic longitudes of equinoxes and solstices

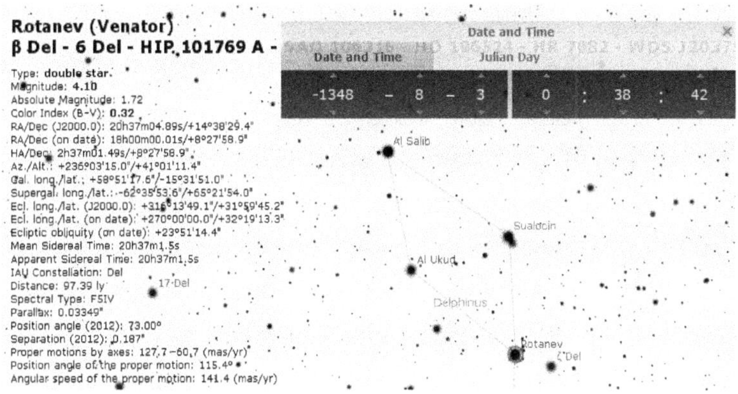

Figure 12.3: The date of winter solstice at the yogatārā of Śraviṣṭhā nakṣatra assuming Rotanev (β Delphini) as the yogatārā of Śraviṣṭhā nakṣatra

With the assumption of given coordinates as polar coordinates, the ecliptic longitude of the yogatārā of Śraviṣṭhā has been converted to 296° 15′ from 290° given in the text. This makes the yogatārā of Śraviṣṭhā nearly 3° inside the Śraviṣṭhā nakṣatra, while the evidence is clearly the opposite. Sūrya Siddhānta explicitly mentions that the yogatārā of Śraviṣṭhā nakṣatra was outside the Śraviṣṭhā nakṣatra. According to Sūrya Siddhānta (8.1-9), the yogatārā of Śraviṣṭhā nakṣatra was at the junction of 3rd and 4th quarter of Śravaṇa nakṣatra. Table 5.1 shows that the span of Śraviṣṭhā nakṣatra was between 293° 20′ to 306° 40′, while its yogatārā had the longitude of 290° 0′. Thus when the beginning of Śraviṣṭhā nakṣatra had a longitude of 270°, its yogatārā had the longitude of 266° 40′. Figure 12.4 shows that the assumed yogatārā of Śraviṣṭhā, Rotanev (β Delphini), had a longitude of 266° 40′ in circa 1590 BCE. This is clearly outside the accepted dates of the composition of Vedāṅga Jyotiṣa. However the date of the composition of Vedāṅga Jyotiṣa was even earlier as discussed below.

Śraviṣṭhā nakṣatra has four stars as shown in Figure 12.5. The details of these stars are given in Table 12.1. Currently, Rotanev (β Del) is the accepted yogatārā of Śraviṣṭhā nakṣatra. However, the yogatārā of Śraviṣṭhā nakṣatra should be Al Salib (γ2 Del) as described in Chapter 9. The star Al Salib (γ2 Del) has relative longitude of 289° 57′ in the Aświnī-beginning Rohiṇī system, which is an excellent match with the 290° 0′ longitude given in Sūrya Siddhānta. Thus, Al Salib (γ2 Del) has a better claim of being the yogatārā of Dhaniṣṭhā than Rotanev (β Del).

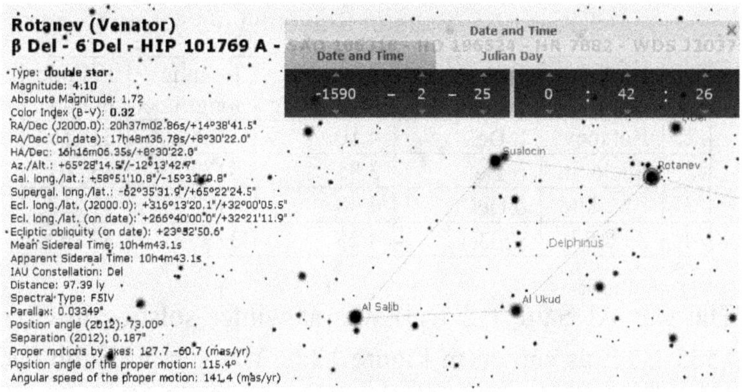

Figure 12.4: The date of the winter solstice at the beginning of Śraviṣṭhā nakṣatra assuming Rotanev (β Delphini) as the yogatārā of Śraviṣṭhā nakṣatra

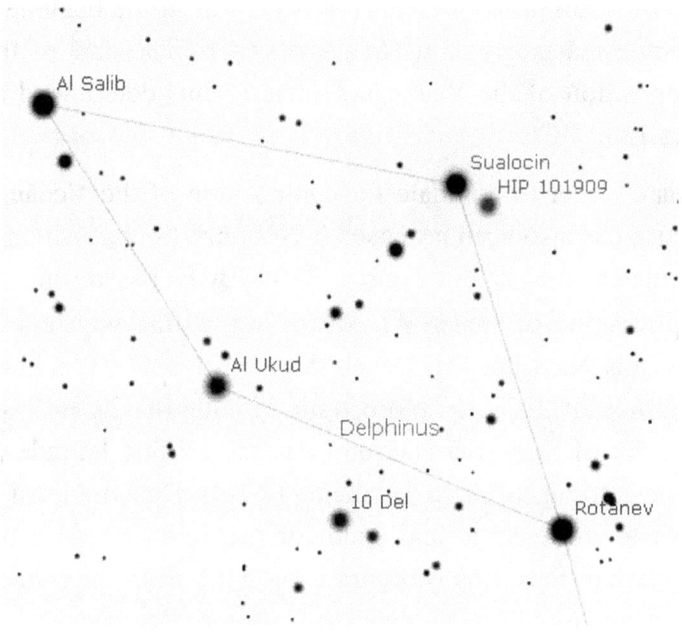

Figure 12.5: Stars in Śraviṣṭhā nakṣatra

Table 12.1: Stars in Śraviṣṭhā nakṣatra

	Proper name	Bayer designation	Apparent Magnitude	Ecliptic longitude	Ecliptic latitude
1.	Rotanev	β Del	4.10	316° 20′	31° 55′
2.	Sualocin	α Del	3.85	317° 23′	33° 01′
3.	Al Ukud	δ Del	4.40	318° 07′	31° 57′
4.	Al Salib	γ2 Del	4.25	319° 22′	32° 42′

The star Al Salib (γ2 Del) was at winter solstice in circa 1585 BCE as shown in Figure 12.6. The star Al Salib had ecliptic longitude of 266° 40′ in circa 1830 BCE as shown in Figure 12.7. As discussed in this chapter earlier, when winter solstice was at the beginning of Śraviṣṭhā nakṣatra, its yogatārā had an ecliptic longitude of 266° 40′. Based on the identification of Al Salib (γ2 Del) as the yogatārā of Śraviṣṭhā nakṣatra, the winter solstice was at the beginning of Śraviṣṭhā nakṣatra in circa 1830 BCE. The date of the composition of the Vedāṅga Jyotiṣa is thus determined as circa 1830 BCE.

A date closer to this date for composition of the Vedāṅga Jyotiṣa has also been proposed by Narahari Achar, who has calculated the date of circa 1800 BCE based on the identification of Deneb Algedi (δ Cap) as the yogatārā of Śraviṣṭhā Nakṣatra [3]. Deneb Algedi has J2000.0 ecliptic longitude of 323° 33′ and ecliptic latitude of -2° 36′. The yogatārā of Śraviṣṭhā Nakṣatra has the ecliptic latitude of 36° according to Sūrya Siddhānta (8.1-9). Deneb Algedi is close to the ecliptic and south of the ecliptic, while the yogatārā of Śraviṣṭhā Nakṣatra is far north from the ecliptic according to Sūrya Siddhānta. Thus the identification of Deneb Algedi as the yogatārā of Śraviṣṭhā Nakṣatra is without merit.

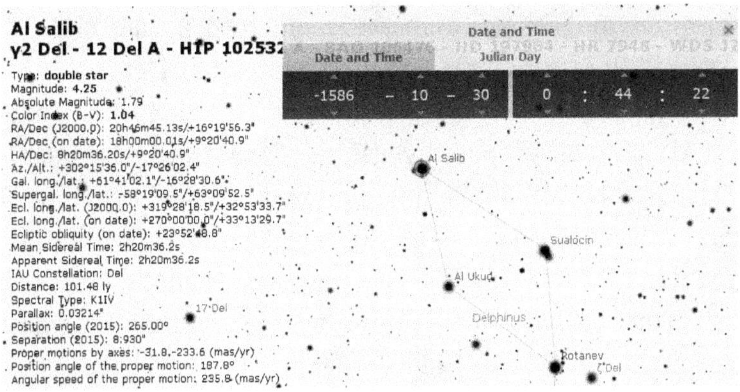

Figure 12.6: The date of winter solstice at the yogatārā of Śraviṣṭhā nakṣatra assuming Al Salib (γ2 Delphini) as the yogatārā of Śraviṣṭhā nakṣatra

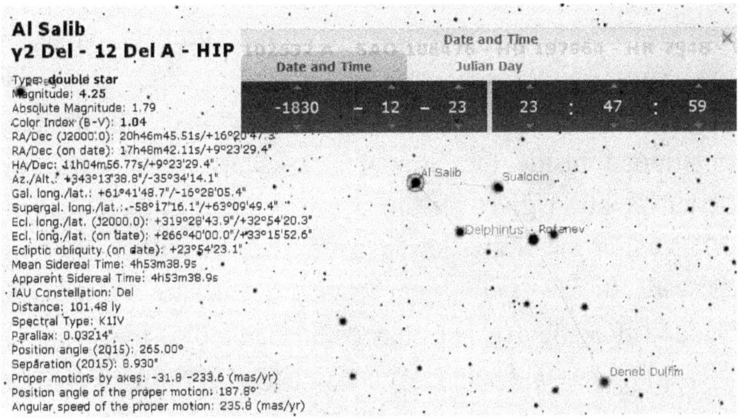

Figure 12.7: The date of the winter solstice at the beginning of Śraviṣṭhā nakṣatra assuming Al Salib (γ2 Delphini) as the yogatārā of Śraviṣṭhā nakṣatra

The reason Narahari Achar has obtained a date close to 1800 BCE is due to the ecliptic longitude of Deneb Algedi being approximately 4° greater than the ecliptic longitude of Al Salib, the proposed yogatārā of Śraviṣṭhā nakṣatra. As discussed above, the ecliptic longitude of the beginning of Śraviṣṭhā nakṣatra is 3° 20′ greater than the ecliptic longitude of its yogatārā. This makes the ecliptic longitude of Deneb Algedi very close to the ecliptic longitude specified in Sūrya Siddhānta (8.1-9). Thus the close matching of the date of composition of the Vedāṅga Jyotiṣa derived by Narahari Achar with this work is fortuitous and not based on in-depth analysis of the textual data.

To put things in perspective, the date of the composition of the Vedāṅga Jyotiṣa has been brought forward by up to 650 years based on two factors. First, the longitude of the yogatārā of Śraviṣṭhā nakṣatra has been increased from 290° to 296° 15′ by asserting that the coordinates given in the Sūrya Siddhānta are polar coordinates. Thus the yogatārā of Śraviṣṭhā nakṣatra has been artificially positioned inside the Śraviṣṭhā nakṣatra by 2° 55′. The actual position given in the Sūrya Siddhānta (8.1-9) is 3° 20′ outside the Śraviṣṭhā nakṣatra. This results in pushing forward the beginning of Śraviṣṭhā nakṣatra by 6° 15′. Second, Rotanev (β Del) has been selected as the yogatārā instead of Al Salib (γ2 Del). The difference in their ecliptic longitudes is approximately 3°. In effect, the beginning of Śraviṣṭhā nakṣatra has been pushed forward by 9° 15′. Each degree amounts to pushing forward the Indian history by 72 years. Thus 9° 15′ is equivalent to pushing Indian history forward by approximately 666 years. This is roughly the time difference between the beginnings of Mauryan era and

Gupta era. This is also roughly the time difference between Cyrus Śaka era (550 BCE) and Śālivāhana Śaka era (78 CE). Thus we see that both historical information and astronomical information have been misinterpreted to give the impression that they corroborate each other. The result of this deliberate manipulation is that one of the greatest astronomers of India, Varāhamihira, has been pushed forward in time by more than six centuries. A detailed analysis of the literary and astronomical information is needed to assign Varāhamihira his correct time in history.

Notes:

1. Kuppanna Sastry (1985): 13-15.
2. Venkatachelam (1953): 50.
3. Narahari Achar (2000).

"When you change the way you look at things, the things you look at change."
—Max Planck

13. The Dating of Varāhamihira

According to the Indian tradition, Varāhamihira was one of the nine gems in the court of Emperor Vikramāditya. Vikrama era named after Emperor Vikramāditya has its zero point in 57 BCE. Modern history denies the existence of Emperor Vikramāditya in 57 BCE and has placed Varāhamihira in sixth century CE, whose time was in 1st century BCE if Indian tradition is to be believed. The reason for this disagreement lies in the interpretation of Śaka era used by Varāhamihira.

13.1 A tale of two eras

Varāhamihira has himself given his date by saying that he wrote Pañchasiddhāntikā in 427 Śaka [1]. Based on the zero point of Śaka era in 78 CE, Varāhamihira wrote Pañchasiddhāntikā in 505 CE. However, Varāhamihira has defined the Śaka era used by him in Bṛhat Saṃhitā, which is reproduced below:

> āsanmaghāsu munayaḥ śāsati pṛithvīn yudhiṣṭhire nṛipatau
> ṣaḍdvikapañcadviyutaḥ śakakālastasya rājyasya. [2]

English translation of this verse has been provided by Alexander Cunningham in 1883 CE [3] as follows:

> The seven seers were in Maghā when king Yudhiṣṭhira ruled the earth, and the period of that king is 2526 years before the Śaka era. [3]

The verse gives the information that the difference between the zero points of Yudhiṣṭhira era and Śaka era was 2,526 years. This information can have two different interpretations, and the chronology of Indian history depends on which interpretation is correct.

Alexander Cunningham interpreted this verse as defining the time of King Yudhiṣṭhira and calculated the time of Yudhiṣṭhira as 2448 BCE, which is 2526 years before 78 CE. However, Indian tradition places the time of Yudhiṣṭhira close to 3102 BCE, the traditionally accepted date of the Mahābhārata war in which Yudhiṣṭhira participated. Venkatachelam calculated that Varāhamihira used the Śaka era with zero point in 550 BCE and proposed that Cyrus the Great was the Śaka king after whom Varāhamihira has named his Śaka era:

> So the Śaka Era related in the Sloka is neither Vikrama nor Salivahana Era and this fact is approved by all the historians. That is the age of the Persian Emperor, Cyrus, which began in 550 B.C. [4]

The difference between these two Śaka eras is 628 years. As pointed out earlier, this is about the same time span by which the ancient Indian history has been pushed forward to give credit to the Greeks for scientific inventions and take away credit from the ancient Indians.

However, 628 years is a long time period and astronomical observations reported by Varāhamihira can be used to fix his time period. The trouble is that scientists and historians are using the assumed date of Varāhamihira to fix the date of his astronomical observations instead of direct verification.

13.2 Precession of the solstices

Varāhamihira has made the following astronomical observation that can settle the debate about his time period:

> Currently Sun turns southward from the beginning of Karkaṭaka and turns other way from the beginning of Makara. If in the future there is deviation from this, then this should be ascertained by direct observation. [5]

From the wordings of the verse, it is clear that Varāhamihira was using sidereal zodiac and not tropical zodiac. In a tropical zodiac, the statement will always be valid. To fix the time period of Varāhamihira, we need to fix the time when the Sun turned southward from the beginning of Karkaṭaka. Table 13.1 shows the Hindu and western names of zodiac signs. Accordingly, Karka or Karkaṭaka is same as Cancer. The question then is -- when did the Sun turn southward from the beginning of Karkaṭaka (Cancer)?

The answer lies in the phenomenon of the "Precession of the Equinoxes", which is the wobbling of the earth's axis. Due to precession, the position of sun among the background of the stars during equinoxes keeps changing. The Sun returns to the same position in about 26,000 years.

Table 13.1: Correspondence of Zodiac signs to Rāśi

	Zodiac sign	Rāśi
1.	Aries	Meṣa
2.	Taurus	Vṛṣabha
3.	Gemini	Mithuna
4.	Cancer	Karka
5.	Leo	Siṃha
6.	Virgo	Kanyā
7.	Libra	Tulā
8.	Scorpio	Vṛścika
9.	Sagittarius	Dhanus
10.	Capricornus	Makara
11.	Aquarius	Kumbha
12.	Pisces	Mīna

As the zodiac has 12 signs, it takes approximately 2160 years to transit through one zodiac sign. The time when the Sun turned southward from the beginning of Karkaṭaka (Cancer) can be calculated by knowing the position of the summer solstice at some other time. Professor James B. Kaler of the University of Illinois has given the following date for the transition of the summer solstice from Gemini to Taurus:

> As a result of precession, around 1990 the Summer Solstice crossed the modern boundary from Gemini to Taurus, which now technically holds the point. Because the Summer Solstice is closer to the classic figure of Gemini than it is to that of Taurus, and since Gemini (along with Pisces, Libra, and Sagittarius) quarters the ecliptic, Gemini is still traditionally taken as the Solstice's celestial home. [6]

Since the Sun turned southward from the beginning of Gemini around 1990 CE and it takes about 2160 years to transit through one zodiac sign, we can calculate backwards and get the approximate dates for previous transitions as shown in Figure 13.1. Thus it was around 170 BCE that the Sun turned southward from the beginning of Karkaṭaka (Cancer). This matches well with the traditional date of Varāhamihira. Placing Varāhamihira in sixth century CE essentially means that the boundaries of sidereal Hindu and western zodiacs do not match and have a difference of about 10° as shown in Figure 13.2.

It is currently believed that the concept of zodiac signs was borrowed by Hindus from Greeks. Then it begs the question as to why the boundaries do not match between Hindu and Western zodiac signs. In fact, the boundaries of sidereal Hindu and Western zodiac signs match very well. Varāhamihira has given information on matching the sidereal Hindu zodiac signs or rāśis with nakṣatras, which provides a means to date his astronomical observations with accuracy.

13.3 Boundaries of rāśis and nakṣatras

Varāhamihira has stated that the beginnings of Meṣa rāśi and Aświnī nakṣatra are the same [7]. As the zodiac signs divide the ecliptic in 12, each zodiac sign spans for 30 degrees. Since there are 27 nakṣatras, each nakṣatra has a span of 13° 20′. Based on this, the correspondence of rāśis and nakṣatras is shown in Figure 13.3. It shows the boundaries of Western zodiac signs and rāśis as identical, which can be corroborated as follows.

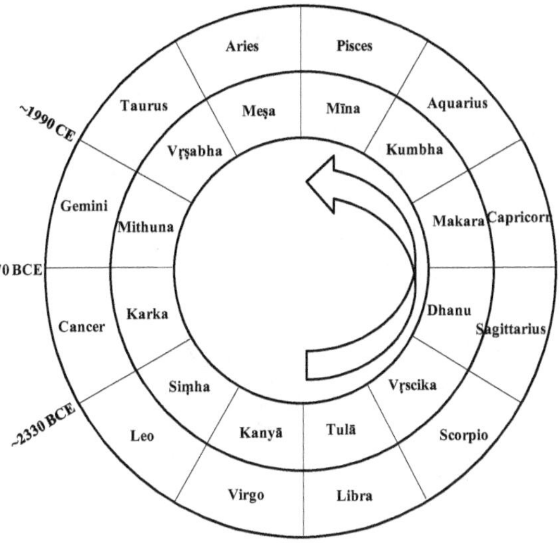

Figure 13.1: Transition of summer solstice into new Zodiac due to precession

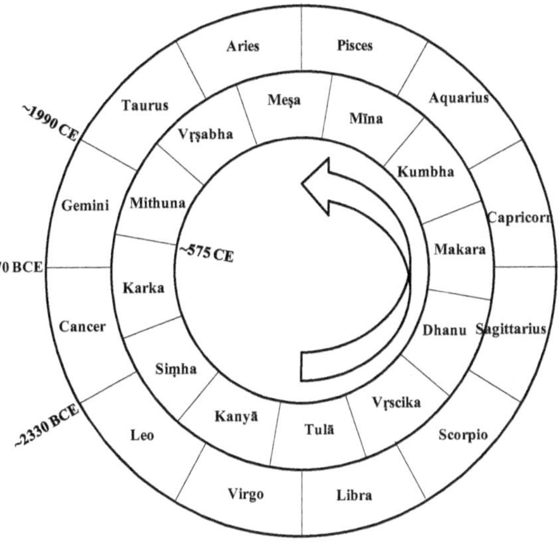

Figure 13.2: Current understanding of the transition of summer solstice into new Zodiac due to precession

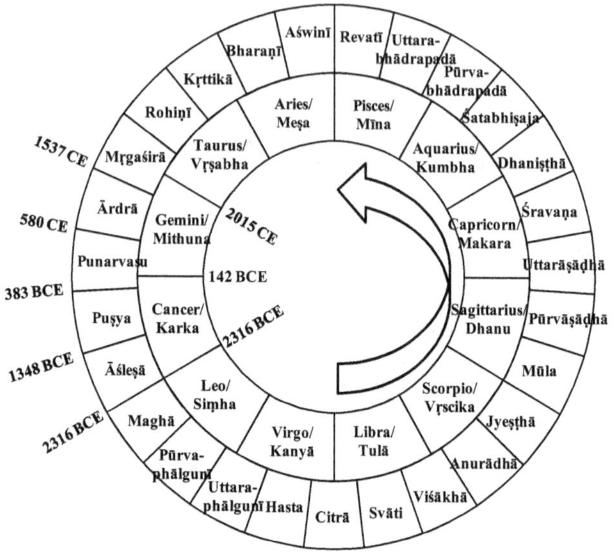

Figure 13.3: Date of summer solstice at nakṣatra and rāśi boundaries

In a paper published in 1904 on the passage of the vernal equinox from Taurus into Aries, it is stated that the longitude of α Arietis for 1900 was 36° 15′ and the longitude of first point of Aries was 28° 22′ [8]. This means that α Arietis was 7° 53′ or approximately 8° from the boundary of Aries. In Chapter 9, it was shown that Hamal or α Arietis is the yogatārā of Aświnī, whose longitude is given as 8° in Sūrya Siddhānta. This means that α Arietis was at 8° from the beginning of Aries as well as the beginning of Aświnī. Since the beginning of Aświnī nakṣatra coincides with the beginning of Meṣa rāśi, it shows that the beginning of Aries coincides with the beginning of Meṣa rāśi.

With the determination of rāśi boundaries and their correspondence with nakṣatra boundaries, precise dates of the position of summer solstice at rāśi and nakṣatra boundaries can be determined as shown in Figure 13.3. Rohiṇī system has been used for these calculations as Hamal or α Arietis, the yogatārā of Aświnī, has a longitude of 8° in this system. The position of the yogatārā of Rohiṇī, Aldebaran, used for these calculations is shown in Table 13.2.

Table 13.2: Summer solstice at rāśi and nakṣatra boundaries in Rohiṇī system

Position	Duration	Reference
Beginning of Maghā	2316 BCE	Aldebaran at 10° 0′
Beginning of Siṃha	2316 BCE	Aldebaran at 10° 0′
Beginning of Āśleṣā	1348 BCE	Aldebaran at 23° 20′
Beginning of Puṣya	383 BCE	Aldebaran at 36° 40′
Beginning of Karka	142 BCE	Aldebaran at 40° 0′
Beginning of Punarvasu	580 CE	Aldebaran at 50° 0′
Beginning of Ārdrā	1537 CE	Aldebaran at 63° 20′
Beginning of Mithuna	2015 CE	Aldebaran at 70° 0′

Varāhamihira has also stated that the Sun changed its course from the middle of Āśleṣā earlier, but now that takes place in Punarvasu [9]. If we look at Figure 13.3, we find that the beginning of Cancer falls in Punarvasu (2/3rd from the beginning of Punarvasu), so Varāhamihira's statements are consistent.

From Figure 13.3 and Table 13.2, it is clear that the astronomical observation made by Varāhamihira took place around 140 BCE. The dating of Varāhamihira to sixth century CE is thus ruled out. Varāhamihira should be dated using the Cyrus Śaka era of 550 BCE. Varāhamihira wrote Pañchasiddhāntikā in 427 Śaka or 123 BCE. One major problem needs to be resolved before Varāhamihira could be dated to 2nd to 1st century BCE. This is the problem of Varāhamihira quoting Āryabhaṭa in Pañchasiddhāntikā [10].

13.4 Āryabhaṭa I or Āryabhaṭa II

Āryabhaṭa has given information about his birth in Āryabhaṭīya [11]. The information has been translated to mean that Āryabhaṭa was 23 years old in the year 3600 of the Kali era. Counting Kali era from the zero point in 3102 BCE, it is claimed that Āryabhaṭa wrote the book Āryabhaṭīya in 499 CE and was born in 476 CE. Varāhamihira could not have lived in 2nd to 1st century BCE, if he has quoted Āryabhaṭa who lived in 5th to 6th century CE. First of all, there is doubt whether Āryabhaṭa was born in 476 CE, 499 CE or 522 CE. Haridatta in 689 CE has interpreted the verse to mean that Āryabhaṭa was born in 499 CE and wrote Āryabhaṭīya in 522 CE, when he was 23 years old [12]. Commentator Someśvara (11th century CE) has interpreted the verse to mean that Āryabhaṭa was born 23 years after 3600 years of Kali era had passed [13].

> "Strange to say, commentator Someśvara understands the verse to mean that 3623 years had elapsed of the Kali Yuga at the birth of Āryabhaṭa."

If this interpretation is correct, then Varāhamihira could not have referred to the Āryabhaṭa of sixth century CE born after he wrote Pañchasiddhāntikā in 505 CE. Even if Āryabhaṭa wrote his famous book Āryabhaṭīya in 499 CE, it is unlikely that he would have been referred by Varāhamihira. Āryabhaṭa lived in Kusumpura, which modern historians identify with Patna in Bihar, while Varāhamihira wrote his treatise in far away Ujjain. It is unlikely that Āryabhaṭa would have become so famous in a mere six years to be quoted by Varāhamihira in an age 1500 years ago, when information travelled much more slowly and it took much longer to build one's reputation. Varāhamihira was referring to an earlier Āryabhaṭa, and not the author of Āryabhaṭīya. It becomes clear from the two quotes by Al-Biruni (11th century):

> In the book of Āryabhaṭa of Kusumapura we read that the mountain Meru is in Himavant, the cold zone, not higher than a yojana. In the translation, however, it has been rendered so as to express that it is not higher than Himavant by more than a yojana. This author is not identical with the elder Āryabhaṭa, but he belongs to his followers, for he quotes him and follows his example. I do not know which of these two namesakes is meant by Balabhadra. [14]

> I have not been able to find anything of the books of Āryabhaṭa. All I know of him I know through the quotations from him given by Brahmagupta. The latter says in a treatise called Critical Research on the Basis of the Canons, that according to Āryabhaṭa the sum of the days of a caturyuga is 1377,917,500, i.e. 300 days less than according to Pulisa. Therefore Āryabhaṭa would give to a kalpa 1,590,540,840,000 days. According to Āryabhaṭa and Pulisa, the kalpa and caturyuga begin with midnight which follows after the day the beginning

> of which is the beginning of the kalpa, according to Brahmagupta. Āryabhaṭa of Kusumapura, who belongs to the school of the elder Āryabhaṭa, says in a small book of his on Al-ntf (?), that '1008 caturyugas are one day of Brahman. The first half of 504 caturyugas is called utsarpini, during which the sun is ascending, and the second half is called avasarpini, during which the sun is descending. The midst of this period is called sama, i.e. equality, for it is the midst of the day, and the two ends are called durtama (?).' [15]

These two statements clearly show that there was another Āryabhaṭa before the Āryabhaṭa of Kusumapura born in 476 CE or 499 CE. Varāhamihira has referred to an earlier Āryabhaṭa about whom not much is known at this point. Also, Varāhamihira could not be contemporary of Āryabhaṭa of six century as that Āryabhaṭa had fixed the date of Mahabharata war in 3102 BCE. Being his contemporary, Varāhamihira could not have fixed the date of Mahabharata war around 2448 BCE.

To put things in perspective, the date of the Varāhamihira has been brought forward by over six centuries based on calculating the date of Varāhamihira from Śaka era with zero point in 78 CE instead of the Cyrus Śaka era with zero point in 550 BCE. The dating of Varāhamihira and Sūrya Siddhānta to 6th century CE was part of the plan by colonial historians to deny the ancient Indians the credit they deserved and show that Hindu astronomers borrowed the basics of astronomy from foreigners such as Babylonians and Greeks. After all, colonized couldn't be allowed to hold their heads high, lest the colonizers lose the legitimacy to rule the colonized. It is with this mindset that important data pinpointing the location of the centre of Vedic astronomy has been proclaimed as borrowed by Indians from Babylonians.

Notes:

1. Pañchasiddhāntikā 1.8.
2. Bṛhat Saṃhitā 13.3.
3. Cunningham (1883): 9.
4. Venkatachelam (1953): 50.
5. Bṛhat Saṃhitā 3.2.
6. http://stars.astro.illinois.edu/celsph.html.
7. Bṛhat Jātaka 1.4.
8. Maunder and Maunder (1904).
9. Pañchasiddhāntikā 3.20-22.
10. Pañchasiddhāntikā 15.20.
11. Āryabhaṭīya, Kālakriyāpāda, Verse 10.
12. Sarma (2001).
13. Dāji (1865). Quote on page 406.
14. Sachau (1910): 246.
15. Sachau (1910): 370-371.

"A people without the knowledge of their past history, origin and culture is like a tree without roots."
—Marcus Garvey

14. The Centre of Vedic Astronomy

According to Vedāṅga Jyotiṣa there are 30 muhūrtas in a day and night (Ṛk Vedāṅga Jyotiṣa 16, Yajus Vedāṅga Jyotiṣa 38) and during the course of the year days and night increase or decrease by a maximum of 6 muhūrtas (Ṛk Vedāṅga Jyotiṣa 7, Yajus Vedāṅga Jyotiṣa 8). Thus the ratio of daylight duration to night duration was 1.5 (3:2) on summer solstice and the ratio of night duration to daylight duration was 1.5 (3:2) on winter solstice. The ratio of longest daylight duration to shortest night duration is a function of latitude, and this information can be used to locate the place where this observation was made.

Colonial era scholars were motivated to proclaim that Indians were borrowers either from Babylonians or from Greeks without providing any proof. Their opinions still continue to be the official position as described below:

> Characteristic of the middle period is the fact that the longest day is considered independent of the geographic latitude and that the ratio of the longest day to the shortest day is taken to be 3:2. This ratio corresponds to a geographic latitude of almost 34°, too high for all parts

> of India except the northwestern corner. THIBAUT mentions that this ratio might be of Babylonian origin but considers this very unlikely because textual evidence was not available. In the meantime, however, KUGLER discovered that the ratio 3:2 occurs in Babylonian cuneiform texts of the Seleucid period. This, coupled with the fact that the ratio of 3:2 was considered in antiquity characteristic for the climate of Babylon, makes it very plausible that the ratio was taken over by the Hindus without correction. [1]
>
> The ratio 3:2 used by the Indians, however, was commonly utilized in all Babylonian astronomical texts after ca 700 B.C. This tradition must surely be the source of the Sanskrit texts under discussion, and provide us with a terminus post quem for those texts. [2]

Based on the Nakṣatra positions given in the Vedāṅga Jyotiṣa, it is currently dated to 1150 BCE to 1400 BCE [3]. It was shown in the Chapter 12 that the actual date of the composition of the Vedāṅga Jyotiṣa is even earlier, closer to 1830 BCE. Since Vedāṅga Jyotiṣa is over 450 years older than Babylonian texts even by the most conservative estimate, it is more likely that the ratio of 1.5 was borrowed by Babylonians from Indians. The idea that this ratio was borrowed by Indians is based on the wrong assumption that there is no prominent place in India where this ratio is valid. Even in the quote above by Schmidt [1], it is said that this ratio is valid for north western part of the then united India. This fact is conveniently ignored to proclaim that the ratio of 3:2 was borrowed from Babylonian astronomers. Kuppanna Sastry in his translation of Vedāṅga Jyotiṣa has also noted that the ratio of 1.5 refers to 35 degrees latitude in the extreme north of India [3]. Sharma and Lishk have also argued against the foreign influence on Indian astronomy and proposed that the ratio 3:2 fits the region of Gandhāra as well and was discovered independently.

> Besides, the simplicity of the relation between the ratio 3:2 and 183 days (half the annual course of the Sun) suggests that the Jainas might have searched for a standard place like Gandhāra where a simple relation of this order holds good. … Gandhāra had been a renowned seat of ancient Indian culture, and no abode of any mythological creatures. As Gandhāra and Babylon are situated on latitudes very close to each other, the ratio 3:2 might have been found independently in these two places. [4]

Gandhāra was a kingdom in ancient India. Its most important cities were Puruṣapura (current Peshawar), Puṣkalāvatī (current Charsadda) and Takṣaśilā (current Taxila). The identification of Gandhāra fits the ratio 3:2 well, however, Gandhāra was a wide region. Sharma and Lishk specify a ratio of 1.42 for Gandhāra, but do not specify exactly where in Gandhāra this ratio holds good. It can be shown using modern astronomical calculations that the ratio of 3:2 fits the location of Taxila exactly, which has a latitude of 33.74° and longitude of 72.80°.

The duration of daylight and night for any day of the year at any location in the world can be obtained from U.S. Naval Observatory website [5]. Table 14.1 shows the duration of daylight and night on the 21st day of each month at Taxila. This data is converted into the ratios of the duration of daylight to night and vice-versa in Table 14.2. The data is shown graphically in Figures 14.1 and 14.2.

Table 14.1: Duration of daylight and night at Taxila

		Daylight		Night	
Month	Date	Hour	Minute	Hour	Minute
1	21 January, 2017	10	17	13	43
2	21 February, 2017	11	12	12	48
3	21 March, 2017	12	10	11	50
4	21 April, 2017	13	13	10	47
5	21 May, 2017	14	3	9	57
6	**21 June, 2017**	**14**	**24**	**9**	**36**
7	21 July, 2017	14	4	9	56
8	21 August, 2017	13	13	10	47
9	21 September, 2017	12	11	11	49
10	21 October, 2017	11	10	12	50
11	21 November, 2017	10	17	13	43
12	**21 December, 2017**	**9**	**55**	**14**	**5**

Table 14.2: The ratio of daylight duration to night duration (D/N) and vice versa (N/D) at Taxila

Month	Date	D/N	N/D
1	21 January, 2017	0.75	1.33
2	21 February, 2017	0.88	1.14
3	21 March, 2017	1.03	0.97
4	21 April, 2017	1.23	0.82
5	21 May, 2017	1.41	0.71
6	**21 June, 2017**	**1.50**	0.67
7	21 July, 2017	1.42	0.71
8	21 August, 2017	1.23	0.82
9	21 September, 2017	1.03	0.97
10	21 October, 2017	0.87	1.15
11	21 November, 2017	0.75	1.33
12	**21 December, 2017**	0.70	**1.42**

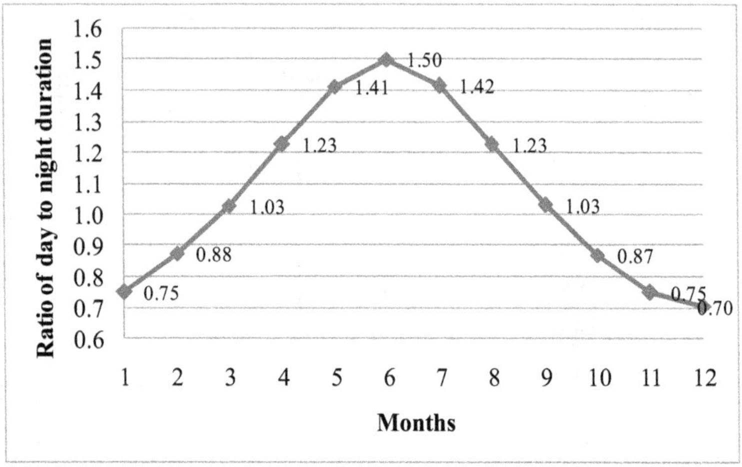

Figure 14.1: The ratio of daylight duration to night duration at Taxila

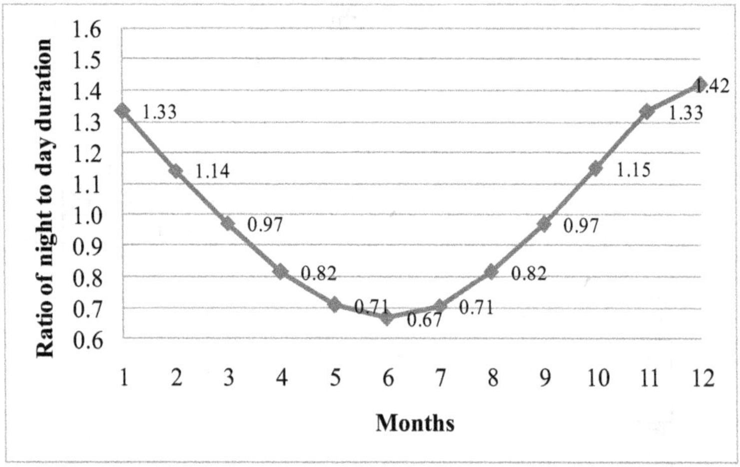

Figure 14.2: The ratio of night duration to daylight duration at Taxila

The choice of 21st day of each month is based on the fact that summer and winter solstices as well as spring and fall equinoxes take place around 21st of the respective months. Months 1 to 12 in these Tables and Figures refer to months January to December respectively.

From Table 14.2 and Figure 14.1, it can be seen that the ratio of daylight duration to night duration at Taxila on summer solstice is 1.5 and matches exactly with the ratio given in Vedāṅga Jyotiṣa. From Table 14.2 and Figure 14.2, it can be seen that the ratio of night duration to daylight duration on winter solstice at Taxila is 1.42 and close to the value of 1.5 given in Vedāṅga Jyotiṣa. Since the data has been obtained for the year 2017, it is natural to ask whether this data is applicable to the observations made during 2nd millennium BCE. The duration of daylight is a function of latitude and tilt of earth's axis to the ecliptic. For fixed latitude, the duration over long time will only depend on earth's tilt. Earth's tilt is currently approximately 23.5 degrees. According to NASA Earth Observatory website [6], earth's tilt changes from 22.1 to 24.5 degrees over a period of 40,000 years. Over a 40,000 year cycle, earth's tilt changes by only 2.4 degrees. Thus the earth's tilt could have differed by no more than 0.24 degrees from the present during the time of Vedāṅga Jyotiṣa and the duration of daylight would not have been significantly different from the values we have now.

From the discussion above, it is clear that there is a specific location in ancient India, namely Taxila, where the observations about the ratios of day and night durations are satisfied. Taxila was the most ancient centre of learning in India with the famous Takṣaśilā University located there.

The discovery of Taxila as the location where Vedāṅga Jyotiṣa was composed shows that Taxila was the centre of Vedic astronomy. It is important to note that Taxila is located in the area where Indus Valley Civilization flourished. This discovery points to the continuity of ancient Indian Civilization. A date of 1830 BCE for the composition of the Vedāṅga Jyotiṣa also shows that the Vedāṅga Jyotiṣa was written when Indus Valley Civilization was coming to an end and refutes the Aryan Invasion Theory. It is important to understand that the Aryan Invasion Theory was invented to legitimize the rule of invaders and colonize the mind of the conquered. The purpose behind sliding the ancient Indian history forward by six and half centuries was to show that Indians were never advanced. It was to show that Indians lacked creativity and were borrowers from the Greeks. Currently, Purāṇas are considered to have been finalized during the reign of Imperial Guptas, who are dated to fourth century. If the Imperial Guptas started their reign in fourth century BCE, then the astronomical information contained in the Purāṇas was discovered by Indians over six centuries prior to the currently accepted date. This will significantly alter the narrative of Indian astronomers borrowing information from Greeks. The truth is that ancient Indian civilization was a knowledge-based civilization and was far advanced than the contemporary civilizations. The flow of knowledge was from India to the west, but to establish it the correct chronological framework of history needs to be developed. I hope that this work will motivate the seekers of truth to investigate further and give the ancient Indians their due place in history.

Notes:

1. Schmidt (1944).
2. Pingree (1973).
3. Kuppanna Sastry (1985): 13-14.
4. Sharma and Lishk (1979).
5. http://aa.usno.navy.mil/data/docs/Dur_OneYear.php.
6. https://earthobservatory.nasa.gov/Features/Milankovitch/milankovitch_2.php.

Bibliography

Abhyankar, K.D. (1991). Misidentification of some Indian Nakṣatras, Indian Journal of History of Science, 26.1: 1-10.

Basham, A.L. (1982). Aśoka and Buddhism – A Re-examination. The Journal of the International Association of Buddhistic Studies, 5 (1): 131-143.

Bhandarkar, R.G. (1872). A Tāmba Patra or Ancient Copper Plate Grant from Kāthiāwād. Indian Antiquary, 1: 14-18.

Bühler, G. (1888). Gurjara Inscriptions, No. III: A New Grant of Dadda II or Prasantaraga. The Indian Antiquary, 17: 183-201.

Burgess, E. (1860). Translation of the Surya-Siddhanta: A Text-Book of Hindu astronomy, with notes, and an appendix. Journal of the American Oriental Society, 6: 141-498.

Charpentier, J. (1931). Antiochus, King of the Yavanas. Bulletin of the School of Oriental Studies, 6 (2): 303-321.

CSIR, (1955). Report of the Calendar Reform Committee. New Delhi, India: Council of Scientific and Industrial Research.

Cunningham, A. (1883). Book of Indian Eras, with Tables for calculating Indian Dates. London, UK: Thacker, Spink and Co.

Dāji, B. (1865). Brief notes on the age and authenticity of the work of Āryabhaṭa, Varāhamihira, Brahmagupta, Bhaṭṭotpala, and Bhāskarāchārya. Journal of the Royal

Asiatic Society of Great Britain & Ireland, New Series, Volume the First: 392-418.

Davids, T. W. R. (1877). International Numismata Orientalia: On the Ancient Coins and Measures of Ceylon. London, UK: Trubner & Co.

Dietz, S. (1995). The Dating of the Historical Buddha in the History of Western Scholarship up to 1980. In "When Did the Buddha Live? The Controversy on the Dating of the Historical Buddha", edited by Heinz Bechert, Delhi, India: Sri Satguru Publications.

Falk, H. (2001). The yuga of Sphujidhvaja and the era of the Kuṣāṇas. Silk Road Art and Archaeology, 7: 121-136.

Falk, H. and Bennett, C. (2009). Macedonian Intercalary Months and the Era of Azes. Acta Orientalia, 70: 197-216.

Fergusson, J. (1870). Art. II. – On Indian Chronology. Journal of the Royal Asiatic Society of Great Britain and Ireland, New Series, Volume the Fourth: 81-137.

Fleet, J.F. (1888). Corpus Inscriptionum Indicarum, Vol. III: Inscriptions of the Early Guptas. Calcutta, India: Government of India, Central Publications Branch.

Goyala, S. (1986). Harṣa Śīlāditya (in Hindi). Meerut, U.P., India: Kusumāñjali Prakāśana.

Goyala, S. (1987a). Gupta Sāmrājya kā Itihāsa (in Hindi). Meerut, U.P., India: Kusumāñjali Prakāśana.

Goyala, S. (1987b). Samudragupta Parākramāṅka (in Hindi). Meerut, U.P., India: Kusumāñjali Prakāśana.

Hamilton, H.C. (1892). The Geography of Strabo. Volume 1. London, UK: George Bell and Sons.

Hultzsch, E. (1914). The Date of Aśoka. Journal of the Royal Asiatic Society of Great Britain and Ireland, October: 943-951.

Hultzsch, E. (1925). Corpus Inscriptionum Indicarum, Vol. I: Inscriptions of Asoka. New Edition. Oxford, UK: Printed for the Government of India at the Clarendon Press.

Jayaswal, K.P. (1934). An Imperial History of India in a Sanskrit Text. Revised by Rahula Sankrityayana. Lahore, United India: Motilal Banarsi Dass.

Jones, W. (1793). The Tenth Anniversary Discourse. Asiatick Researches or Transactions of the Society Instituted in Bengal, 4: xii-xiv.

Kaye, G.R. (1924). Hindu Astronomy, Calcutta: Government of India, Central Publication Branch.

Keith, A.B. (1914). The Veda of the Black Yajus School entitled Taittiriya Sanhita. Cambridge, Massachusetts, USA: Harvard University Press.

Kuppanna Sastry, T.S. (1985). Vedāṅga Jyotiṣa of Lagadha in its Ṛk and Yajus recensions. New Delhi: Indian National Science Academy.

Legge, J. (1886). A Record of Buddhistic Kingdoms, Being an Account by the Chinese Monk Fa-Hien of His Travels in India and Ceylon (A.D. 399-414) in Search of the Buddhistic Books of Discipline. Oxford, UK: Clarendon Press.

Majumdar, R. C. and Altekar, A. S. (editors). (1967). The Vakataka-Gupta Age. Delhi, India: Motilal Banarasidass.

Malla, K. P. (2005). Mānadeva Samvat: An investigation into an Historical Fraud. Contributions to Nepalese Studies, 32 (1), 1-49.

Maunder, E.W. and Maunder, A.S.D. (1904). Note on the date of the passage of the vernal equinox from Taurus into Aries. Monthly Notices of the Royal Astronomical Society, Volume 64, Issue 5, 11 March 1904: 488-506.

McCrindle, J.W. (1877). Ancient India as Described by Megasthenes and Arrian. London, UK: Trubner & Co.

McCrindle, J.W. (1893). The Invasion of India by Alexander the Great. Westminster, UK: Archibald Constable and Co.

McCrindle, J.W. (1901). Ancient India as Described in Classical Literature. Westminster, UK: Archibald Constable and Co.

Mirashi, V.V. (editor). (1955a). Corpus Inscriptionum Indicarum, Vol. IV: Inscriptions of the Kalachuri-Chedi era. Part 1. New Delhi, India: Archaeological Survey of India.

Mirashi, V.V. (editor). (1955b). Corpus Inscriptionum Indicarum, Vol. IV: Inscriptions of the Kalachuri-Chedi era. Part 2. New Delhi, India: Archaeological Survey of India.

Mukhopadhyaya, S. (1963). The Aśokavadana. Delhi, India: Sahitya Akademi.

Narahari Achar, B.N. (2000). A case for revisiting the date of Vedāṅga Jyotiṣa. Indian Journal of History of Science, 35(3): 173-183.

Pingree, D. (1973). The Mesopotamian origin of early Indian mathematical astronomy. Journal of the history of astronomy, 4: 1-12.

Pingree, D. and Morrissey, P. (1989). On the identification of the "Yogatārās" of the Indian Nakṣatras. Journal for the History of Astronomy, 20(2): 99-119.

Prinsep, J. (1837). Interpretation of the most ancient of the inscriptions on the pillar called the lat of Feroz Shah, near Delhi, and of the Allahabad, Radhia and Mattiah pillar, or lat, inscriptions which agree therewith. Journal of Royal Asiatic Society of Bengal, July: 566-609.

Prinsep, J. (1838a). Discovery of the name of Antiochus the Great, in two of the edicts of Aśoka, king of India. Journal of Royal Asiatic Society of Bengal, February: 156-167.

Prinsep, J. (1838b). On the edicts of Piyadasi, or Aśoka, the Buddhist monarch of India, preserved on the Girnar rock in the Gujerat peninsula, and on the Dhauli rock in Cuttack; with the discovery of Ptolemy's name therein. Journal of Royal Asiatic Society of Bengal, March: 219-282.

Rao, N. L., (1931-32). Gokak plates of Dejja-Maharaja. Epigraphia Indica, 21: 289-292.

Roy, R.R.M. (2015a). India before Alexander: A new Chronology. Mississauga, Ontario, Canada: Mount Meru Publishing.
https://www.amazon.com/gp/product/B018HDUELG/ref=dbs_a_def_rwt_bibl_vppi_i1.

Roy, R.R.M. (2015b). India after Alexander: The Age of Vikramādityas. Mississauga, Ontario, Canada: Mount Meru Publishing.

https://www.amazon.com/gp/product/B018IKTFM2/ref=dbs_a_def_rwt_bibl_vppi_i3.

Roy, R.R.M. (2015c). India after Vikramāditya: The Melting Pot. Mississauga, Ontario, Canada: Mount Meru Publishing.
https://www.amazon.com/gp/product/B018JP6RK4/ref=dbs_a_def_rwt_bibl_vppi_i2.

Roy, R.R.M. (2019). Sidereal Ecliptic Coordinate System of Sūrya Siddhānta, Indian Journal of History of Science, 54(3): 286-303.
https://insa.nic.in/writereaddata/UpLoadedFiles/IJHS/Vol54_3_2019__Art03.pdf.

Sachau, E. C. (1910). Alberuni's India. Vol. 2. London, UK: Kegan Paul, Trench, Trubner & Co. Ltd.

Sagar, K. C. (1992). Foreign Influence on Ancient India. New Delhi, India: Northern Book Centre.

Sarma, K.V. (2001). Āryabhaṭa: His name, time and provenance. Indian Journal of History of Science, 36(3-4): 105-115.

Śarmā, R. (1995). Bhārata ke Prāchina Nagaroṃ kā Patana (in Hindi). New Delhi, India: Rājakamala Prakāśana.

Schmidt, O. H. (1944). The Computation of the Length of Daylight in Hindu Astronomy. Isis, 35(3): 205-211.

Sethna, K.D. (1989). Ancient India in a New Light. New Delhi, India: Aditya Prakashana.

Sharma, S.D. and Lishk, S.S. (1979). Length of the day in Jaina astronomy. Centaurus, 22(3):165-176.

Siṃha, U.N. (1986). Sūrya Siddhānta (with Hindi translation and extensive introduction). Kolkata: Śrīmatī Savitrī Devī Bāgaḍia Trust.

Sircar, D.C. (1965). Indian Epigraphy, First edition. Delhi, India: Motilal Banarsidass.

Srinivasan, D. M. (2007). On the Cusp of an Era. Leiden, Netherlands: Brill.

Śrīvāstava, M.P. (1982). Sūrya Siddhānta (with scientific commentary). Allahabad, India: Dr. Ratnakumarī Swādhyāya Saṃsthāna.

Strong, J. S. (1989). The legend of King Aśoka: A study and translation of the Aśokavadana. Delhi, India: Motilal Banarsidass.

Thapar, R. (2013). The Past before us. Cambridge, Massachusetts, USA: Harvard University Press.

Thomas, E. (1858). Essays on Indian Antiquities, historic, Numismatic, and Paleographic of the late James Prinsep to which are added his Useful Tables, illustrative of Indian history, chronology, modern coinage, weights, measures, etc. Volume II. London, UK: John Murray.

Vassilkov, Y. V. (1997-98). On the meaning of the names Aśoka and Piyadasi. Indologica Taurinensia, 23-24: 441-457.

Venkatachar, M.A. (2014). Yogataaras, their misidentifications and corrective measures. https://astronomyrevisitedbymav.files.wordpress.com/2017/06/yoga-final-with-figures-date.pdf.

Venkatachelam, K. (1953). The plot in Indian Chronology. Ghandhinagara/Vijayawada, India: Bharata Charitra Bhaskara.

Whitney, W.D. (1905). Atharva-Veda Saṃhitā, Second Half. Cambridge, Massachusetts, USA: Harvard University Press.

Wilson, H. H. (1850). On the Rock Inscription of Kapur Di Giri, Dhauli, and Girnar. The Journal of the Royal Asiatic Society of Great Britain and Ireland, Volume the Twelfth: 153-251.

Index

Also by Mount Meru Publishing

Author: Dr. Raja Ram Mohan Roy

1. Vedic Physics: Scientific Origin of Hinduism
2. India before Alexander: A New Chronology
3. India after Alexander: The Age of Vikramādityas
4. India after Vikramāditya: The Melting Pot
5. Zero point of Jain Astronomy: The Origin of Malava Era

Author: Professor Subhash Kak

1. The Circle of Memory: An Autobiography
2. Matter and Mind: The Vaisheshika Sutra of Kanada
3. Arrival and Exile: Selected Poems
4. Computation in Ancient India
5. Mind and Self: Patanjali's Yoga Sutra and Modern Science
6. The Nature of Physical Reality (Third Edition)

Author: Dr. Dilip Amin

1. Interfaith Marriage: Share and Respect with Equality

Author: Professor Ramesh Rao

1. The Election that Shaped Gujarat and Narendra Modi's Rise to National Stardom

About the Author

Dr. Raja Ram Mohan Roy studied at Netarhat Residential School in then state of Bihar, now in Jharkhand. He earned his undergraduate degree in Metallurgical Engineering from Indian Institute of Technology, Kanpur and Ph.D. in Materials Science and Engineering from The Ohio State University, USA. He moved to Canada as a Postdoctoral Fellow. Raja has conducted research and development in the areas of Extractive Metallurgy and Materials Processing for twenty years. He has co-authored 40 research papers that have been published in peer-reviewed journals and proceedings of international symposia. He has co-edited the book "Innovative Process Development in Metallurgical Industry."

Raja has always had a fascination for ancient Indian civilization. He is the author of Vedic Physics: Scientific Origin of Hinduism, India before Alexander: A New Chronology, India after Alexander: The Age of Vikramādityas, India after Vikramāditya: The Melting Pot, and Zero Point of Jain Astronomy: The Origin of Malava Era. Through his writings, Raja hopes to contribute towards the continuity and understanding of his civilization and, in the Indic tradition, repay the debt to his ancestors for their contributions and their sacrifices.

www.ingramcontent.com/pod-product-compliance
Lightning Source LLC
Chambersburg PA
CBHW060018210326
41520CB00009B/927
* 9 7 8 1 9 8 8 2 0 7 2 4 7 *